REPOWER
High School STEM

REPOWER
High School STEM

21st-Century STEM Education
Problems Cannot Be Solved
With a 19th-Century Academic
Structure

Kenneth M. Chapman

Cardinal Workforce Developers, LLC

First Printing, 2022
Ruther Glen, Virginia, USA
Cover: Agricultural power sketches by Karen Nolan

In an era when:

- A friend using a donor's heart continues to contribute to the community by operating a hardware store that manages to provide off-the-shelf plumbing repair supplies not available at the expansive big box stores of national chains.
- Millions of people routinely travel to destinations thousands of miles distant in a few hours.
- I can make pies with fresh strawberries when there are no leaves on deciduous trees.
- Anyone can watch the tragedy of wildfires in Australia in real-time.
- A soldier (safeguards require several soldiers) pushing buttons can kill millions of humans within minutes.
- A research center involving thousands of scientists and engineers from across the world with communications delivered in microseconds can operate a complex device several miles in diameter to identify the existence of a particle of subatomic dimensions.

We continue to use a STEM education structure developed for an era when:

- The first organ transplant was nearly four decades in the future. (https://www.google.com/search?q=organ+transplant+history&oq=organ+transplant+his-

tory&aqs=chrome..69i57j0l7.6259j0j8&sour-
ceid=chrome&ie=UTF-8)

- Many students rode horses to schools not within walking distance.
- Most families subsisted in winter on food harvested from their gardens and preserved.
- Commercial radio broadcasts would become available several years in the future.
- Using recently developed technology, a soldier could kill a few hundred attackers in a day.
- Most scientists either worked alone with only local resources, or as members of teams working while widely separated with communications delivered in months.

Artifacts and procedures continue to be added to an archaic science education arrangement straining to survive, while potential applications of new technology and organizations are ignored. Perhaps publication of this Model STEM System (MSS) will stimulate some quantum changes in STEM education.

The Limits of Operating Structures

Like many cats, Ace takes great interest in the laser pointer, and wants desperately to capture the red dot. However, despite his inherent intelligence and his extensive skills in running, stopping, turning, pouncing, and ambushing, he consistently fails in his attempts. Why? One reason is that his physical form — in other words, his operating structure — is incompatible with his objective.

Similarly, the current structure of traditional high school STEM gives us a no-possible-win situation. The 19th-century high school structure imposed on STEM education fails to facilitate the efforts of many dedicated teachers trying to prepare their students for 21st-century conditions that require creativity, critical thinking, creative problem-solving, and collaboration skills.

Principal Components of the Model STEM System

One 3- or 4-year STEM (Science, Technology, Engineering, and Mathematics) Course: Replaces the traditional biology, chemistry, and physics sequence. See Chapter 3.

STEM Database Center (SDC): Repository of STEM education standards; CSS connections; project materials; and assessment items (diagnostic, formative, and summative). At teacher direction, it may include school curricula, test construction and analysis; as well as rec-ommendations for instructing individual students based on assessments and student history. *National, probably with satellites.* See Chapter 8.

Catalyst STEM Specialist (CSS): Post-secondary school STEM specialists, ranging from college students to professionals in trades and scientists and engineers in industry and research organizations. See Chapter 6.

STEM Service Center (SSC): A regional unit serving many schools with Catalyst STEM Teachers (CSTs), professional development, and support materials. See Chapter 7.

Teacher Teams: A school's STEM teaching staffs that have direct contact with students, manage teaching elements, design curricula, provide lead teachers for discrete projects, implement mini-courses, and provide subject experts as needed. Ideally, the teacher team is a four-member unit consisting of one specialist each for biology, chemistry, physics, and engineering. The teacher team may be supplemented with STEM Catalyst Teachers (SCTs) as needed. See Chapter 6.

Projects and Mini-courses: The principal components of the STEM course. See Chapters 4 and 5.

CONTENTS

CONTENTS

ACKNOWLEDGMENTS

First, I wish to thank my wife, Ginny, and our family for supporting my ideas for STEM education and converting the ideas into book form. While the work took place in and near my home, the work hours were numerous and often led to very early morning hours of writing with naps during the day.

Throughout my career, many students, educators, technicians, scientists, and engineers gave me opportunities to observe attempts to improve secondary and post-secondary education. Edward Fleckenstein, an electrical engineer, enabled me to administer programs and teach chemistry and chemical engineering in two colleges. Dr. Moses Passer hired me twice to work at the American Chemical Society. He imposed few constraints, which allowed me to remain immersed in science education at a national level and, starting in 1967, to observe efforts to improve high school science education and attract underrepresented minorities to chemistry. Janice Carneal, Harold Stills, Brent Miller, and Carolyn Williamson gave me incredible support while I taught high school science classes.

I had the privilege from 1970-72 to work with Dr. Robert Pecsok and a team of 23 chemists and two chemical engineers, representing both academe and industry, to write a textbook series for the chemistry core of chemical technology programs in the Chemical Technician Curriculum Project or ChemTeC. The team members were stars in their own work arenas. However, during a 10-week period in the summer of 1970, they merged in an ideal way, with excellent support staff, to create nearly 800 pages of edited text, with two high-quality volumes of more than 100 pages each returned from the printer for classroom use and the remainder prepared for printing. The team achieved this production without the use of computers and with copiers of limited capability. The ChemTeC team was a model for an ideal team.

Eric Stewart, through his copyediting and science writing experience, contributed vast improvements to the original manuscript. However, he is exempted from all errors caused by my additions and changes.

A Challenge to the Reader

You, Dear Reader, are challenged to answer the question:
> *"What would I design for high school STEM education
> if I could start with a clean slate?"*

Your answer may differ significantly from the one presented in this book. Your unique experiences will flavor your answer strongly. However, suppose your answer looked much like the traditional science program most U.S. high school students are experiencing in the early twenty-first century. In that case, I would be disappointed – and so should you.

The traditional structure for education in a few STEM subjects, developed between 1892 and 1919 during a period of astounding industrialization in the United States, was created for students either living in rural areas or having intimate knowledge of rural life. High school-age students brought into their classrooms problem-solving experiences; many had direct knowledge of the physical and mental skills needed for building things, growing biological products, and animal husbandry. Such daily experiences enabled many of these students to observe and speculate about natural phenomena. Teachers of science could give answers and assess the information absorbed. Most of today's students, in contrast, bring radically different sets of skills and backgrounds to the high school classroom.

Even in the early 20th century, high school students expected to encounter a very different world in which to survive than did their parents. The career advisors of teenage students in 1920 could not predict opportunities in computer systems, high-speed worldwide communication devices fitting easily into a pocket, humans walking on the moon, or meat grown in a laboratory. Neither can today's advisors predict the changes for which their advisees must prepare. What skills and knowledge are worth teaching in high school STEM classes? Those who have created national and state STEM education standards have struggled with the question and provided their best answers for the present, and perhaps a few years into the future. As an instance of how radically things change, note that since the Next Generation Science Standards were published (2013), the definition of the kilogram has changed!

The Proposal: Create a structure for high school STEM education that offers all students an opportunity for success to replace the traditional science courses that encourage avoidance or failure for many students.

A Personal Perspective: I believe high school STEM education needs to change by a significant quantum jump instead of relying solely on evolutionary changes, as it has since the late 19th century. Drop-outs as well as graduates should leave high school well-qualified for further education or work in STEM occupations; even students who believe they have no talent for STEM still must be readied for life as citizens in a technological world. Many policy makers seeking to lead the nation through the Covid-19 epidemic have demonstrated an inadequate understanding of STEM philosophy and knowledge, leaving many citizens confused and distrustful of "experts." STEM education also provides excellent opportunities for students to develop non-technical skills valuable for life and work in non-STEM fields.

Learning for STEM now often starts in kindergarten and continues throughout life. A severe chokepoint for the STEM workforce occurs when students pass through their high school years. The same choke point often adversely affects many rising citizens. The inadequacy of much high school STEM education compromises national survival, the work-life balance of many individuals, and choices made daily by every citizen. During my professional life in STEM workforce development, I formed these perspectives and observations of my fellow citizens, both directly and in the public discourse.

All the stakeholders of STEM should have roles in high school education. Most education committees in government agencies and

STEM membership organizations have industry representatives. However, these industry stakeholders are usually few in number, and their perspectives seldom receive much attention. Their representation should be increased, and their voices need to be heard and respected.

Many STEM specialists with rich career experiences in industry and research could be prepared to support high school teachers and catalyze classroom activities in ways that conserve their time and maximize their impact. Connecting high school STEM teachers and STEM specialists is difficult, even in industrialized locales. Fortunately, modern communications and database structures offer great potential for facilitating connectivity between these critical groups.

Most educators continue to use the term "science" when they actually mean to refer to the broader scope of "STEM." Meanwhile, in traditional high school classrooms, the same term is usually used in a narrow context of seeking only **the** correct answer. This restriction fails to address the nuance of how applied science is used in many situations, where several alternative answers may satisfy a single question. As in engineering design, the best solution depends upon many variables and may change as conditions shift. Educators should consider the "science" in STEM to include "applied science," thus giving teachers and counselors more varied opportunities to work with students.

American high schools vary from well-supported institutions serving thousands of students with STEM faculties with excellent preparation, to small schools serving a few students with one part-time STEM teacher of limited background in most of the subjects they teach. Usually, the latter schools also lack sufficient STEM facilities and supplies. Consequently, most school support systems do not enable all high school STEM students to receive equivalent instruction.

STEM Is Vital to Both the Nation and Citizens: STEM education is critically important to the economic viability of the United States, as well as to the well-being of American citizens. STEM education is ideal for stimulating the creativity of students. Critical thinking and creativity are individual traits that should be developed, and also linked to respect for the thoughts and perspectives of others. STEM classrooms can enable students to learn STEM subjects while also developing their talents for working effectively with others for future success in the workplace, communities, and families.

A very popular TED Talk was delivered in 2006 by Sir Ken Robinson on schools killing creativity.[1] For many years, Robinson encouraged replacing an outmoded industrial education system with a highly personalized approach. This undertaking would apply the remarkable technological and professional resources that could engage all students and enable them to prepare more thoroughly for 21st-century challenges for work and citizenship.[2] He called for a revolution in schools and gave examples of excellent efforts to change; but he did not, so far as I am aware, suggest specifically what that revolution would mean to reform for high school STEM classroom operations.

Nick Hanauer, a venture capitalist and philanthropist, has concluded that economic inequities must be addressed before better school experiences become available to most students. In a July 2019 article[3], he includes a quotation from economist Lawrence Mishel about the plight of economically poor children: "... that children who frequently change schools due to poor housing; have little help with homework; have few role models of success; have more exposure to lead and asbestos; have an untreated vision, ear, dental, or other health problem; ... and live in a chaotic and frequently unsafe environment."[4] Political and employment issues that impact education and equity cannot be solved by schools alone. However, STEM operations in disadvantaged high schools can be repowered to give students more equitable opportunities for high-quality education and recognize potential occupations in their adult lives.

As an engineer, I was pleased that the Next Generation Science Standards (NGSS) recognized and included engineering and technology when published in 2013. Some students become bored with situations where single science concepts lead only to one correct answer. Many students respond more enthusiastically to opportunities to do practical problem-solving where they can be more creative. Engineers typically explore multiple alternatives, each one a potentially acceptable answer, for solving a problem, and then choose the option that best suits current and/or foreseeable conditions.

Ten years of high school science teaching in my first retirement emphasized the extreme need for more help for teachers. Every new report and editorial addressing STEM education seems to carry additional expectations of teachers. New aids provided through computer and audiovisual systems are often intended to help teachers. However, they usually require adding more demands of teacher time for learning to use and then maintaining accuracy and currency as the devices and software are applied. Teachers need and deserve real help, and technology needs to be harnessed more effectively for STEM education. When teachers are supported, students benefit.

The Model STEM System (MSS) described in this book would support teachers by applying computer technology in new ways and enabling STEM specialists to participate directly in classroom settings. It suggests a resource to augment both teaching and material issues when a school has deficiencies. It emphasizes a structure and methodology to help students improve creativity, critical thinking, and teamwork skills. This MSS seeks to stimulate thinking about a revolution in STEM teaching, rather than just attempting to layer improvements on top of a 19th-century course structure. That structure, in my view, was better suited to prepare students to produce high-volume consumer products on an assembly line than tackling twenty-first-century problems and opportunities. Extensive research must precede any implementation of the MSS to improve and test its components and convince naysayers of its viability. Both national security and the well-being of our citizens are at stake.

Most high schools' traditional approach to STEM education addresses a few of the problems identified by Robinson and Hanauer as teachers try to help students build bridges to more education. Some school systems engage local employers to find new sources of classroom support and work experiences, or motivate students with other unique and locally created methods.

The MSS suggested here, on the other hand, directly attacks some of the problems associated with the absence of role models and inadequate relevancy. It presents a way for the non-academic STEM community to engage catalytically in STEM education in all high schools at low cost. Inaction will continue to be much more costly (in terms of lost opportunities for individuals, employers, and society) than the expenses of a research-proven revolution in STEM education structure.

Reform Without Improvement: "Reform education!" was a rallying cry heard all over Washington, DC, when I arrived in the city to work full-time for the American Chemical Society. More than 53 years later, multiple commission reports, studies of student testing averages, international comparisons of education results, and complaints from both colleges and employers continue to support the need to "reform education." Many brilliant and well-meaning educators – classroom teachers, professors, representatives of stakeholder organizations, government leaders, and philanthropists – have dedicated much of their lives to improving high school STEM education to little effect when comparative testing and post-secondary educators and employer criticisms are the measures. Perhaps the whole structure of high school STEM education needs to be revolutionized. Theodore R. Sizer, a 20th-century reformer, suggested that one has to change everything to make a fundamental change in education.

STEM Education at Its Best and Worst: At one extreme, I had the privilege to serve on the committee that founded the Thomas Jefferson School of Science and Technology in Fairfax County, Virginia, which continues to rank at or near the top of most national high school rankings. At the other extreme, the academic leader of a small private school asked me to help with a student revolt against their chemistry teacher. The students had correctly ascertained the school's only STEM teacher knew little about chemistry. When the 18-year-old school moved to a new physical site, its accumulated high school science supplies and equipment fit comfortably in a four cubic-foot cardboard box. The Thomas Jefferson School of Science and Technology has served by challenging a small cadre of students selected carefully from a large contingent of candidates with rigorous courses often taught by teachers with doctoral degrees in discipline areas of STEM. The latter school made a solid commitment to improve its STEM education and now a group of well-qualified science teachers is dedicated to helping students with varied academic abilities maximize their progress.

National efforts to improve high school STEM education, led by highly competent leaders from STEM fields and STEM education, have occurred frequently throughout my professional life. Yet, there seems to have been meager improvement in student performance, and only a tiny fraction of students leave high school with excellent STEM credentials. Most of their fellow graduates are not ready for the next level of STEM education, and many of them leave post-high school STEM studies and training as failures. Even during high school, many students become STEM dropouts. Far too many students living in underprivileged communities are denied the learning opportunities required for furthering their preparation for employment in STEM fields.

The Tradition: Since the late 19th century, the accepted instructional structure throughout the United States has been a single-subject teacher meeting with a group of students for 45-50 minutes, five days a week for an academic year. In many school systems outside privileged suburbs and urban areas, STEM teachers have a limited ability to give their students continuous access to good role models with up-to-date insights into STEM workplaces and processes. For many classrooms, the traditional structure of STEM education mitigates against providing students access to the best learning opportunities that the broad world of STEM work offers. Of course, any interaction with working STEM personnel at any level must accompany minimal impositions for employers, employees, and entrepreneurs.

Publishers have made significant investments in creating arrays of carefully written textbooks supported with exquisite media resources.

Similarly, equipment purveyors have created highly innovative devices and computer programs for high school laboratories to mimic those used in STEM workplaces.

The national investment in time, personnel, and physical structures for teaching STEM in high schools, and the vested interests of many teachers, offer enormous inertial forces to oppose any significant structural changes. However, many other segments in modern life with similar investments *have* changed. In one lifetime, for example, American farmers moved from using horses as their principal power source to highly automated tractors. Likewise, transportation moved from horse-drawn buggies and smoky locomotives to comfortable and dependable automobiles and, for long-distance travel, airplanes. In the same lifetime, many facets of life shifted from paper and pencil to computers. Why should the high school STEM education structure remain stagnant when it develops the characteristics of the personnel who drive all these other changes?

Reconsidering STEM Education: An invitation to participate in a meeting to design new curricula and a new building in a neighboring county came to me in late 2015 from a parent of two of my high school students. Sheryl "Ginger" Martin was a retired U.S. Navy officer who at that time supervised several R&D teams in naval research that included personnel ranging from custodians to Ph.D. scientists.

Although Ginger and I had had several discussions about the urgency of critical workforce issues related to high school education and STEM curricula, I was skeptical that any meeting purporting to discuss the complex issues of curricular change and basic building design would give much attention to the content and strategies of teaching. I decided to accept the invitation, both out of respect for Ginger, and based on many years of personal observations from workforce development and high school classroom teaching, a few of which were:

- After several years of experience teaching the introductory chemistry course to new college students, I was tempted to drop the prerequisite of high school chemistry. Most of my students had many fundamental misconceptions about chemistry that interfered with their learning, leading to numerous dropouts.

- After a three-year hiring freeze at a vast chemical manufacturing site, a company announced only to its current employees that it would accept applications for 75 new positions on one specified day. With no other advertising, the company received over 10,000 in-person applications during that one day. The company eventually hired 100 (about 1 %) of the applicants. Although screened rigorously, one-third of the new hires were dismissed as "untrainable" by the end of the probationary period.
- High school science teachers often are expected to provide information to students about college major selections and workforce opportunities while having little understanding of the broad spectrum of employment opportunities in STEM fields and the variety of post-secondary programs for STEM fields.
- Some high school students were genuinely surprised that some skills learned in their mathematics courses required application in STEM courses.
- A young temporary employee refused to learn basic word processing skills when computers first were used for office operations.
- Both corporate and government-supported projects continually add well-designed and compelling artifacts and software learning materials to high school STEM resources. However, too few students, particularly those in challenging educational situations, can profit from these resources due to cost, time, or teacher discomfort with the technology.

My expectations about the meeting proved correct. However, preparing for the meeting led to my conclusion that curriculum change was not the most fundamental issue; instead, there seemed to be a need to change the structure within which we teach the first courses for high school STEM. Some of the essential knowledge and skills high school students need for life and opportunities in the 21st century that should be addressed effectively through teaching about STEM cannot be fitted into the traditional course structure that has endured for over a century.

Teachers Need Help, Not Admonitions: Frequently, leaders in STEM education describe the needs of students and suggest demonstrably good teaching changes that teachers should support through individual efforts. However, few teachers have the time and breadth of experience to create a whole panoply of STEM content and processes. Teachers are overwhelmed, need direct assistance, and nothing in the present structure provides that help when a classroom door is closed and the teacher and the students are alone.

Put Computers to Use Helping Teachers: The grand promise of computers to help educators is materializing slowly with the development of many useful instructional materials and administrative programs. However, teachers need a computer-based support system to develop and present curricula, identify individual student needs for knowledge and skills, and suggest instructional strategies and materials for individual students. Perhaps the sophistication and power of computer systems and artificial intelligence used routinely for military and business needs can be applied to address teachers' needs.

In the Beginning: In 1892, the Committee of Ten started to work on recommended changes to high school offerings to provide more uniformity among college entrants. At that time, several leading high schools were beginning to separate natural science into chemistry, physics, and several types of biology courses. Charles Eliot, the Committee's chair, was both a chemist and president of Harvard College. He had written previously that "practical end[s] should never be lost sight of by student or teacher in a polytechnic school, and should seldom be thought of or alluded to in a college...."[5] Technology and engineering were not considered appropriate subjects for high schools whose purpose was preparing students for college entrance. Both academe and industry exploit the multidisciplinary character of STEM today. Perhaps today's students would be served better if the three introductory high school biology, chemistry, and physics courses were restructured and "technology and engineering" (as presented by the NGSS) were added.

Stacks of One-room Schoolhouses: Concurrently with integrating the disciplines, the custom of pairing one teacher in a classroom with one group of students needs reexamining. Many STEM teachers serve small schools, or work with schedules that preclude professional exchange with others. Few can find the time to engage with professionals in their primary discipline area, even outside of school hours. Perhaps the classroom model patterned after the one-room schoolhouse for a single discipline should be reconsidered.

A Source for Helping Teachers and Students: The ratio of high school science teachers to scientists and engineers is about 1:50. Adding all the others whose work applies STEM concepts would swell the ratio to over 100 STEM workers for each high school STEM teacher. Perhaps some of these many STEM workers could use modern communications to provide direct classroom/lab assistance to the beleaguered teachers, and catalyze students' understanding of how STEM works beyond high school, while also giving career information.

There Is More to STEM Education than Just STEM: Much rhetoric and efforts by education leaders, standards developers, textbook creators, and concerned business leaders suggest STEM classes should address more than just science and technical content. Critical thinking, problem-solving, and data analysis are considered essential skills. Yet, standardized tests and other pressures encourage many teachers to revert to teaching "safe" science content only. High school graduates' knowledge and skills would be much more valuable if they could practice some of the skills associated with applying STEM in the world outside the high school classroom — skills that receive little attention in current courses.

Projects Versus Lectures: To deliver voluminous information and perspectives, many teachers rely on lecturing. Much evidence suggests lecturing is a poor way to teach anyone other than adults specifically interested in the subject matter. Projects, on the other hand, stimulate student interest and provide an effective conduit for learning both content and skills. However, many STEM teachers limit the scope of or truncate projects, due to inadequate time, resources, knowledge, and skill — even within the closely confined boundaries of activities the teachers themselves created. Perhaps modern communications could be applied to incorporating the expertise of STEM experts and design specialists into many student project team operations.

Current Change Efforts: Many American educators apply their insights and energies to serve high school STEM students better. Many teachers make brilliant changes in their individual classrooms. The principals and staff of some individual schools have made valuable changes to STEM teaching practices. Other reforms are driven across districts by school boards, superintendents, and staff. Some states have instituted admirable reforms by creating employment opportunities for students in STEM fields through state-level leaders and educators at all levels. Reforms driven by these implementers are valuable. However, many students remain untouched by such efforts. The disparity between the students profiting from reform efforts and those who are untouched grows wider each year. New resources and restructuring certain teaching practices could greatly diminish that disparity while further enhancing students' learning.

Most Current STEM Support Work Should Continue: Structural changes in teaching STEM in high schools should not interfere with many efforts already supplementing classroom work. FIRST is a robotics competition that fosters teamwork skills and provides exposure to insights and skills in several essential areas of STEM applications. The Pathways system, promoted by the National Center for College and Career Transition, advocates a broad program for engaging local businesses with students and school staff to connect school learning with local needs and opportunities. Colorado has a state-wide program for placing high school students into meaningful work. The University of Connecticut conducts a five-day course for introducing drug discovery and development to high school students. The American Chemical

Society has operated Project SEED since the late 1960s to place eco-nomically-disadvantaged high school students in college and industrial research laboratories during summer breaks with some opportunities continuing in the school year.

Many other organizations sponsor similar activities to encourage further study and consideration of careers in STEM fields. However, of the approximately 15 million high school students in the United States, many lack the time and resources needed to participate in these valuable programs. As a result, these programs may actually widen the gap be-tween students privileged to study science in exceptional circumstances and those relegated to substandard schools, with limited faculties and few opportunities to encounter supportive STEM specialists.

A Model from Outside the Traditional Education Box: The Model STEM System described in the following chapters is an attempt to demonstrate how some of the problems implied above might be mitigated. This proposed approach requires only minor changes out-side the high school science or STEM classrooms, with four significant exceptions: (1) STEM personnel outside the high school must become involved with student work in classrooms; (2) teachers must become prepared to address team building and operation; (3) STEM teachers must be served by an easily-used large and complex computer center; and (4) support centers for STEM education must become part of the education complex. Doing less denies many students opportunities to improve their futures and erodes the capacities of employers and society generally.

Teachers receive substantial intrinsic rewards when they sense that students are learning deeply from a well-presented lesson. The MSS does not reduce such accolades, although the teacher must share them with others.

After creating a system like MSS, operational costs per student will be significantly higher than for the present high school STEM struc-ture. However, a cost/benefit analysis relative to the nation's economy and security should show a substantial gain. When so many potential improvements are possible in a new system, it is morally reprehensible to deny opportunities for a better future to those trapped in schools that are severely limited by the traditional structure.

The purpose of the MSS is not to seek adoption directly but to stimulate discussion about changing the structure of high school STEM education and the development of implementable designs. Failure to address the intractable problems in the current high school STEM structure will adversely affect the nation's economy, national security, and opportunities for students from diverse populations.

For the origin and development process for the Model STEM System, see Appendices C and D.

Kenneth Chapman

Ruther Glen, Virginia   2022

References

1. Robinson, K., Do Schools Kill Creativity? TED Talk, 2006. https://www.ted.com/talks/sir_ken_robin-son_do_schools_kill_creativity/transcript?language=en (accessed Dec 21, 2021).

2. Robinson, K., *Creative Schools*; Penguin Books: New York, 2015.

3. Hanauer, N., Better Schools Won't Fix America. *The Atlantic*, July 2019. https://www.theatlantic.com/magazine/archive/2019/07/education-isnt-enough/590611/ (accessed Dec 21, 2021).

4. Mishel, L., The Opportunity Dodge. *The American Prospect*, Spring 2015. https://prospect.org/power/opportunity-dodge/ (accessed Dec 21, 2021).

5. Eliot, C., The New Education. *The Atlantic Monthly*, Vol. 23, 1869, p. 203.

INTRODUCTION

Problems and Assumptions

Defining STEM

The opening paragraph of the National Science Teaching Association's 2020 position statement entitled, "STEM Education Teaching and Learning," provides this overview of STEM education:

> "The National Science Teaching Association (NSTA) strongly supports STEM (science, technology, engineering, and mathematics) education that provides students with an interdisciplinary approach to learning. STEM education makes learning 'real' and gives students opportunities to see the connection between the content they are studying and the application of that content in authentic and relevant ways."[6]

In *The Case for STEM Education*, published in 2013,[7] Dr. Rodger Bybee, a contributor to the NSTA position statement, identified nine common perspectives for using the acronym STEM. One of these perspectives serves as the basis for the definition of STEM applied generally to the Model STEM System (MSS) presented in this book:

> *"STEM is a transdisciplinary course (a system) that envelopes all traditional and new science, engineering, and technology disciplines to be presented in a three- or four-year high school period intended to provide learning and skill development*

*opportunities and necessary supportive mathematical practices
ranging from cutting edge research to perspectives pertinent to
technical trades."*

Examples of the possible inclusiveness of STEM for the high school
may encompass: learning about the properties of a soldering material
for joining pipes, developing computer coding skills, and applying the
fundamental conceptions of atomic models to predict molecular structures. The STEM system suggested here would serve as a foundation
for concurrent or further study and skill development as desired for
careers as varied as plumbers, cybersecurity programmers, designers of
complex devices, and researchers expanding knowledge at the cutting
edge of science.

The STEM course must serve two principal audiences. One consists
of all students, as they will become adults and be involved in a technological society that constantly requires decisions involving the science,
engineering, and technology that affect users in their daily lives. Simultaneously, some students need to acquire specific skills and knowledge
that are the foundations for additional study or careers from trades to
esoteric scientific research.

Bybee identifies other common perspectives about STEM, including
its ability to:

- serve as a new name for science (or mathematics)
- incorporate both science and mathematics
- enhance understanding of science by expanding to include technology, engineering, and mathematics as equal components
- add science and mathematics to technology or engineering programs
- emphasize cooperation across science, technology, engineering,
  and mathematics
- apply to combinations of two or three disciplines
- integrate overlapping and sequencing among the science, technology, engineering, and mathematics disciplines

"STEM" may also be used as an adjective (e.g., STEM employment) to reflect the broad range of adult work that requires training or education, from vocational school to post-graduate study.

Hypotheses for Rebuilding the Structure for STEM

The proposed MSS responds to several expectations of its stakeholders. After the MSS has been used by a school long enough to reach a steady-state condition, the following hypotheses apply:

- **Student outcomes upon completion of the STEM system:** By the end of the third year of the MSS, students having achieved average or better performance levels will perform well in all types of career and technical education and AP science courses. These students will demonstrate mastery of content and processes advocated by the current nationally distributed STEM education standards (NGSS in 2021). These students will also possess knowledge and skills of STEM to enable them to participate in a high technology society and be prepared for future success in entry-level college courses in science and engineering.
- **Student receptivity:** Students will respond positively to an educational regimen that utilizes STEM projects as the base for learning the knowledge and skills for varied careers and life readiness.
- **Equity issues:** Students from populations underrepresented in STEM careers will be encouraged to consider preparation for such work when exposed with some regularity to STEM specialists whose backgrounds are similar to their own.
- **Academically-poor students:** Students will respond positively to instruction that builds on what they do understand, rather than upon what a calendar suggests they should comprehend.
- **Academically-strong students:** Students will respond positively to having STEM specialists available regularly to support advanced interests that cannot be satisfied by the teacher team.

- **Employer response:** Business and government agencies will support educational activities that use their resources sparingly in order to promote positive catalytic effects on students and teachers.
- **Teacher response:** Many experienced teachers and most prospective teachers will support a team-oriented environment that focuses on student learning rather than elapsed time.
- **Technology applications:** Computer technology with artificial intelligence will respond to needs for supporting and assisting teachers, not just providing direct instruction and statistical data.
- **Non-academic STEM specialists:** Many scientists, engineers, and other STEM workers at all levels, when prepared and supported, can engage effectively with high school students to facilitate the learning of both knowledge and processes of STEM, as well as related career and future workforce information.

General STEM Objectives

To prepare to model a high school STEM system, this author started with a blank slate. The first step was to identify necessary objectives, along with several major contravening problems. At the same time, the author considered and clarified several assumptions about American education.

Objectives for the high school STEM program: Several outcomes required to address the future needs of students are necessary for any basic high school STEM program. Also, the MSS addressed in this book uses a single course spread over three (3) years, plus an additional fourth-year elective. The course would begin in the first year of high school or 9th grade. Generalized objectives include:

1. Students who earn a better-than-average grade upon completing three years in the MSS and who are interested in pursuing college-level science, engineering, and technology majors should

be able to do so without remediation. These students also should experience success in the elective fourth-year of STEM or AP STEM courses.

2. All students earning a passing grade should succeed in a concurrent or successor career technical education (CTE) or the fourth-year STEM elective. These students also should find success in apprenticeship programs and basic entry-level STEM jobs in all types of businesses.

3. Upon completing three years in the MSS, all students earning a passing grade should know about STEM content relevant to daily life and non-STEM work, and be able to make justifiable decisions about civic matters.

A Few Problems Encountered by Teachers

In the present STEM structure of sequential courses in biology, chemistry, and physics, STEM teachers encounter several difficult, non-behavioral (sometimes intractable) problems. Several are identified below In broad terms in alphabetic order. Feel free to make additions.

Adolescent malleability: High school students have entered the second developmental period of brain plasticity, but do so at different chronological ages. However, most high school STEM classes tend to be taught in lock-step, with some key content addressed only once during a student's high school career.

Applying all STEM resources: Many non-academic STEM experts and specialists have knowledge and skills that would benefit students greatly; traditionally, however, their use is episodic at best, and few students ever receive any classroom exposure to them. Also, materials and services to support student learning activities could be made available externally to the high school at a reasonable cost and level of effort.

Career information: Few STEM teachers have sufficient experience with various STEM fields to guide students through critical career and college major decision-making. One piece of evidence is a recent study of college graduates that found that **61% would change their majors if they could restart their college experience**[8].

Content: STEM is multidisciplinary, yet few STEM teachers command the broad scientific/engineering/technology content needed to guide students in all possible STEM fields. Standards statements usually give wide latitude rather than narrow specifications or guidance. Educators should continually look to the future and not be lulled into complacency by considering only current standards when judging what content to incorporate into curricula. Authors of the NGSS wrote statements that offer broad opportunities for addressing the specific content students likely need to achieve the developers' intent.

Diversity: Underrepresented minorities (URM) comprise too small a portion of the STEM workforce. African-Americans tend to leave college STEM majors in disproportionately high numbers, although the fraction of their high school graduates entering STEM majors is about the same as Caucasians[9].

Experiments and projects: Most high school science experiments are designed to eliminate risk and minimize cost by giving individuals or student teams only one opportunity to collect data to "prove" a scientific principle, make observations, or develop a skill. Projects often feature individual or small group activities with no connection to any STEM workplace or real-world activity. Seldom is a student team with many members enabled to plan and manage their project.

Grading: A single grade fails to distinguish between preparation for future STEM study or careers and a more broad-based understanding of STEM topics necessary for participation as an informed citizen.

Reacting to STEM news: No mechanism exists for helping teachers incorporate new scientific discoveries and technological developments into classrooms rapidly and at a level that is meaningful and motivational to students.

Resources v. advances/local issues: Textbook systems require enormous financial investments and lend themselves to neither addressing local needs/opportunities nor responding to current STEM-related developments of great interest. Yet, these systems often dictate almost everything that happens in classrooms.

Single-discipline classrooms: Most traditional STEM classrooms cannot be converted conveniently for instruction addressing multiple disciplines.

Standards: STEM standards[10] (with accompanying explanations) tend to be extensive in number and length. Most are not prescriptive, either for guiding classroom activities or evaluating individual students' progress.

Student variability: Some students arrive as 9th graders fully capable of addressing demanding STEM studies and with great enthusiasm; many others have inferior basic arithmetic skills and lack any motivation for studying STEM subjects. Some students cannot avail themselves of all that a teacher or the school offers due to requirements of work to support a family, lack of financial resources, family member care requirements, and lack of transportation.

Students do not become team members: Teamwork is a skill expected of almost every employee from the first day on the job. It often is required in college work and other activities starting in the first days of higher education.

Students are given responsibility only for themselves: Many high school students lack opportunities to exercise responsibility, yet most students respond positively when given real responsibility.

Teacher inadequacies: Most STEM teachers have a good command of content for one discipline area. Expecting them to have adequate knowledge of all other STEM fields cheats students.

Time: Many STEM teachers devote almost all waking hours to preparing for classes, presenting the day's subjects, and following up with test grading and reviewing homework after the school day. Often, professional development is pursued both during school sessions and "vacation" periods.

Using students as teachers: Peer teaching always occurs. However, advanced students are usually not available to teach novices. Teaching is a skill starting to be recognized as needed by every employee at every level.

Assumptions Undergirding the Model STEM System

Assumptions underlay every STEM activity and must be clarified as thoroughly as possible to achieve optimal designs. Many engineering students are encouraged to cite their assumptions clearly as they tackle homework problems. Addressing more complex design and problem-solving challenges professionally, they will seek to eliminate as many assumptions as possible.

Creating the MSS started with the identification of several key assumptions that affect instruction in STEM. This list of assumptions is likely incomplete in important content, but does provide some insight into the reasons for the designs that constitute MSS. Readers are encouraged to modify this list and the MSS to reflect their perspectives and develop better STEM education systems. The following list implies no ranking of the assumptions.

General:

- Our nation's economic wellbeing and national security are dependent on the STEM knowledge and skills of a large portion of the population.
- High school STEM education is the last resort many students have for developing knowledge and skills critical for functioning effectively in today's society. Also, it is an essential link for the workforce development chain and the last opportunity for most potential employees and entrepreneurs to prepare for continuing education to embark on STEM careers.
- Well-meaning and well-designed policies, recommendations, and direct contributions of government agencies, membership

organizations, companies, and individuals often fail to affect the many high school students whose contact with the world of STEM is limited severely. Many students from groups under-represented in STEM attend schools with inadequate facilities and staff. Their education then is limited to the intercession of dedicated teachers who are assumed, often incorrectly, to have the expertise in technical content and instructional techniques to develop new instructional designs, materials, and devices to enhance learning.

- A small but vitally important proportion of high school student cohorts seek to enter STEM professional careers.
- With the nation's ratio of births/deaths declining, the U.S. must maximize the portion of its population qualified to become employed in STEM fields and ensure that groups underrepresented in STEM employment are enabled to become participants.
- No mechanism exists that enables high school STEM teachers to effectively use current and recent advances in science, engineering, and technology to enhance instruction.
- Any movement to implement components of this MSS or similar systems must be preceded by proof-of-concept projects, the creation of many student projects with complete documentation and evaluation, and professional development for many practicing STEM teachers.
- Colleges, disciplinary science and engineering professional organizations, museums, and other educational entities must not be required to make very many significant changes in their strategies and activities in support of high school STEM. Colleges will create programs to develop team teaching expertise for both new and practicing STEM teachers.
- Publishers will restructure instructional materials to be less prescriptive and more supportive of informational STEM materials and skill development.

External Support

- Database management techniques and artificial intelligence (AI) provide untapped capabilities: (a) to make educational standards information much more useful and available to teachers and other educators; (b) to provide analysis of student information and assessments to diagnose problems and inadequacies; and (c) to make suggestions to teachers about possible teaching strategies and relevant resources.
- Employers of STEM personnel and entrepreneurial STEM practitioners will accept time-limited and flexible responsibility for improving the education of high school students.
- From maintenance technicians to Nobel laureates, many STEM specialists will accept roles as Catalyst STEM Specialists (CSS) (see Chapter 6).
- CSSs will provide career models, technical information, and project management perspectives otherwise unavailable to students and teachers.
- STEM Service Centers (see Chapter 7) can be adequately staffed and be financially sustainable.

School Personnel

- The American high school STEM education mechanisms can be improved by applying modern technology and creating a system that supports teachers and schools rather than relying on mandates.
- Teachers with expertise in STEM subjects will be willing to be facilitators of learning and forego being the fount of all relevant knowledge and dictator of the classroom in order to work in a flexible instructional setting.

- Teachers will work effectively as team members both to enhance learning opportunities and to serve as role models.
- STEM teachers can be persuaded to forego their isolated classrooms for a team system that mimics standard practices of the STEM world of work.
- Students will respond favorably to projects requiring them to accept increased responsibility for making and acting on decisions.

School Work

- Using student projects based on authentic settings will provide students a base for analyzing data and information, expressing creativity, and learning foundational concepts and skills while presenting a motivational context and structure.
- Using short mini-courses with students grouped homogeneously by their levels of content knowledge and skills will prove possible, efficient, and effective.
- Modern communications and everyday practices can contribute to effective remote instruction engaging CSSs and/or CSTs (see Chapters 4 and 6).
- Extracurricular activities that provide some students insights into the practices of STEM work will continue to provide motivation, skill development, and valuable career information to limited numbers of students.
- Many high schools lack access to facilities and materials adequate for supporting complex student projects in STEM.
- CSTs can provide the necessary support to STEM teacher teams with inadequate resources to address STEM curriculum design and implementation.
- Media-delivered instruction will be necessary for STEM education. STEM teachers can be supported directly with computer analysis of student information and test results.

- The structure of traditional science education courses, which continue to provide the exposure most high school students experience, is over a century old. Employer complaints, post-secondary remediation requirements, and comparisons based on student test results suggest the structure has not functioned adequately for many years.
- High school students respond well to increased responsibilities.
- Students must engage in their learning.
- Some of the standard content of high school STEM instruction has limited value to many students, but many of the processes of STEM have value outside of STEM activities.
- Restructuring high school STEM education should not have any harmful effects on the application of in-class learning techniques or supportive extracurricular efforts.
- Curriculum choices should be determined locally and informed by national, state, and local recommendations and requirements.
- Sequencing may be less critical to STEM instruction than assumed traditionally but sometimes provides the only logical pathway forward.
- Misconceptions usually are more vexing to learning than a void of knowledge.
- Students with more knowledge and skills often are very effective in teaching less-knowledgeable colleagues.

A Brief Relevant History of Instruction in STEM

The Report of the Committee of Ten[11] of the early 1890s led to the current structure of STEM offerings in most American high schools. The committee chair, Dr. Charles Eliot, a chemist and president of Harvard University, was the author of the first textbook in chemistry using the laboratory method.[12]

The Report, described in 1958 as "the most profound ... philosophy of education ever enunciated in America"[13], recommended that science should constitute about 20% of all instruction in high schools, and suggested a sequence of:

Year	Subject
1	Applied Geography
2	Botany or Zoology
3	Physics and a course in Astronomy and Meteorology
4	Chemistry and one semester of Geology or Physiography and a semester of Anatomy, Physiology, and Hygiene

The Report recommended the first algebra course be given in Year 2.

Early in the twentieth century, another national committee recommended consolidation of several courses into one of biology and advocated changes to the order of science courses to:

Year	Subject
1	Biology
2	Chemistry
3	Physics

Schools varied in their implementation, beginning the sequence in the 9th or 10th grade. The courses tended to be taught independently of each other. In large schools, specialists in biology, chemistry, and physics could teach their respective specialties. In smaller schools, fewer teachers would be available; sometimes, the same teacher taught all the science courses.

In 1957, the launch of Sputnik spurred the National Science Foundation to fund several projects to change the direction of high school science education to become more theoretical and laboratory-centered. Writing new textbooks and retrofitting thousands of teachers required several years. Although some new texts were not commercially successful, their influence on all subsequent competitive texts in the same field was significant.

Still, detractors of American education used statistics of test scores to argue that high school science education was still inadequate. Funded and promoted by the federal government, the report titled *A Nation at Risk*[14] challenged all science educators in 1983. The level of enthusiasm for "education reform" made a quantum jump. The report focused on declining average test scores from standardized SAT exams. Sandia Laboratories was charged in 1990 with examining the SAT data in much greater detail than had been done for the report in order to get a clearer picture of the decline in scores. Breaking the data into population segments revealed that each group's test scores were increasing instead of declining. What had happened was that the number of representatives from poorer performing groups taking the SATs was becoming a more significant portion of all the test takers, thus driving down the overall average. However, an increasing number of college students required remedial work before being admitted to science major courses, and employers continued complaining about the inadequacy of new hires.

The result of efforts by President George H.W. Bush and a national governors conference was an Act of Congress in 1994 that established the "Goals 2000: Educate America Act.[15]" Goal 5 of the Act stated that "United States students will be first in the world in mathematics and science achievement." Unfortunately, our nation has fallen far short of that goal: the Program for International Student Assessment reported that for 2018, of 78 countries doing comparative testing, the United States ranked below 36 countries in mathematics and below 17 countries in science.[16]

In the early 1990s, discussions in many different organizations ultimately led to national standards for K-12 science education. Under the National Research Council's (NRC) auspices, the National Science Education Standards were published in 1996[17]. These were superseded by the Next Generation Science Standards that were published in 2013 by NRC[18]. The latter added engineering and technology to the subjects addressed.

The Case for a STEM Foundation for All

STEM satisfies the innate human drive to understand the world around us. It is utilitarian in that scientific principles enable the design, production, and distribution of components needed to preserve and enjoy life. It also provides the foundation for a wide variety of employment, from digging into the ground to increase food production to scanning space to identify objects that might strike Earth catastrophically. Upon completing STEM study in high schools, students should demonstrate they possess skills for teamwork, analysis, critical thinking, communicating effectively, and organizing their further study of STEM content applicable both inside and outside STEM employment.

Students entering high school begin a period of rapid development as they begin spanning the divide between their childhood years of intense curiosity and becoming adults responsible for sustaining themselves, and possibly others as well. These students develop learning habits and perspectives about the world in which they live. They also hone methods of interacting with fellow humans and lay a foundation for responding to change. In this maelstrom of growth and learning, STEM education helps students transition from a generalized response to curiosity to developing and correlating knowledge and skills that lead to understanding and processes of searching for information, critically thinking, and solving problems.

The high school years are the last opportunity to study STEM content and processes for many students. Also, high school STEM instruction provides many students a foundation for further study and careers.

Parallel Programs to Support STEM in High School

MSS focuses on the high school classroom, laboratory instruction, and project management. The need to prepare students better for decision-making about careers and the preparation they may require has spawned many beneficial parallel programs and extracurricular activities. Students find cooperative education, internships, and other work exposures to be valuable experiences. Schools and individual teachers gain helpful support by engaging with local community groups and employers. Competitions in robotics and presentations to non-academic audiences can be life-changing events for many students. However, many students are unable to participate in extracurricular activities. Pathways, a carefully planned career preparation effort, may be broad enough to incorporate MSS into its core.

Successful participation in MSS should prepare students for career and technical education (CTE) and additional STEM courses such as AP specialties and college credit courses as early as the third year. Some students may find a fourth year of MSS valuable.

Conclusions

If high school STEM instructional objectives are to be achieved, many intractable problems that are difficult to address in the traditional structure of the high school courses of biology, chemistry, and physics must be addressed. Considering STEM teaching in a system context such as MSS may help mitigate many of these problems. However, the MSS structure proposed, or any alternative, has several elements that must be subjected to detailed examination and modification, which requires creating materials, classroom trials, and teacher development. There are many assumptions inherent in any undertaking to make significant changes in STEM education. However, changes should support the intrinsic desires of STEM teachers to help as many of their students as possible to become better prepared for addressing unimagined situations in the future. Many institutions –– colleges, high schools, corporations, non-profit organizations, and government agencies –– must find ways to cooperate to maximize the value of high school STEM education.

References

6. NSTA Board of Directors Position Statement, "STEM Education Teaching and Learning." https://www.nsta.org/nstas-official-positions/stem-education-teaching-and-learning (accessed Dec 21, 2021).

7. Bybee, R. *The Case for STEM Education: Challenges and Opportunities*; NSTA Press: Arlington, VA, 2013.

8. Johnson, R. Most College Grads Would Change Majors. Blog post on BestColleges website. https://www.bestcolleges.com/blog/college-graduate-majors-survey/ (accessed Dec 21, 2021).

9. Bauer-Wolf, J. Early Departures. Article on Inside Higher Education website. https://www.insidehighered.com/news/2019/02/26/latinx-black-college-students-leave-stem-majors-more-white-students (accessed Dec 21, 2021).

10. nextgenscience.org. "Read the Standards" web page. https://www.nextgenscience.org/search-standards (accessed Dec 21, 2021).

11. National Education Association. *Report of the Committee of Ten on Secondary School Studies with the Reports of the Conferences Arranged by the Committee*; American Book Company: New York, 1894. https://archive.org/details/reportcommittee00studgoog/page/n8/mode/2up (accessed Feb 8, 2022).

12. Eliot, C.W. and Stoner, F.H. *A Manual of Inorganic Chemistry, Arranged to Facilitate the Experimental Demonstration of the Facts and Principles of the Science*; Ivison, Phinney, Blakeman, and Co.: New York, 1868.

13. Latimer, J.F. *What's Happened in Our High Schools?*, Washington Public Affairs Press: Washington, DC, 1958.

14. National Commission on Excellence in Education. *A Nation at Risk: The Imperative for Educational Reform*; U.S. Department of Educa-

tion: Washington, DC, 1983 https://www.edreform.com/wp-content/uploads/2013/02/A_Nation_At_Risk_1983.pdf (accessed Dec 21, 2021).

15. Goals 2000: Educate America Act. Public Law 103-227, 1994; Code of Federal Regulations, Section 102(5). https://www.govinfo.gov/content/pkg/BILLS-103hr1804enr/pdf/BILLS-103hr1804enr.pdf (accessed Dec 21, 2021).

16. Program for International Student Assessment. PISA 2018 Worldwide Ranking. https://factsmaps.com/pisa-2018-worldwide-ranking-average-score-of-mathematics-science-reading/ (accessed Dec 21, 2021).

17. National Academies of Science, Engineering, and Medicine. *National Science Education Standards*; National Academies Press: Washington, DC, 1996.

18. NGSS Lead States. *Next Generation Science Standards: For States, By State*s. National Academies Press: Washington, DC, 2013.

2 |

Brief Overview of the Model STEM System (MSS)

One Engineer's Perspectives About High School STEM

The end product of MSS, a class of students exiting the STEM course, may be described by considering a spectrum of the objectives for STEM education as shown in Figure 2-1, which considers the extent of formal study of STEM. At one end of the spectrum, some students will engage in no further courses in STEM after high school. Yet, they will become citizens who use technology and potentially participate in policy-making involving STEM issues. At the other end, a few students may pursue doctoral degrees in science or engineering. In between, many students will stop or postpone further study in STEM in order to work in fields that may depend upon STEM concepts and processes. Thus, the high school STEM course must satisfy many needs, provide motivation, be condescending to no student, and be rigorous.

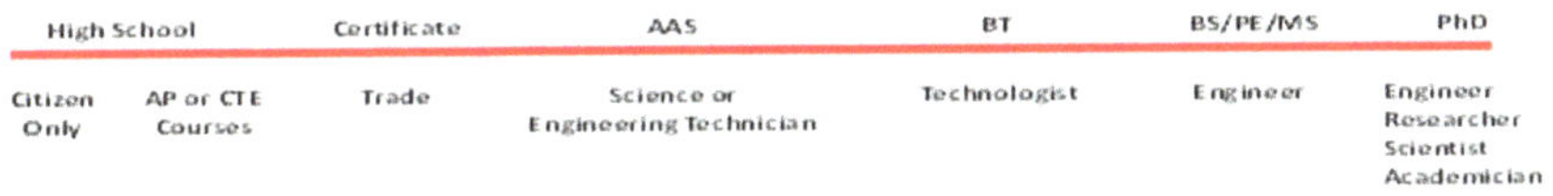

High School	Certificate	AAS	BT	BS/PE/MS	PhD
Citizen Only	AP or CTE Courses	Trade	Science or Engineering Technician	Technologist	Engineer	Engineer Researcher Scientist Academician

Figure 2.1: The Spectrum of Student Objectives

STEM education at the high school level offers many learning incentives, opportunities for exploration or satisfying societal expectations, and academic challenges. STEM satisfies curiosity, explains observations and experiences, stimulates the intellect, correlates multiple disciplines, provides intellectual challenges, and offers careers with good incomes. Yet, far too many students find high school STEM classes boring or overwhelming, and seek to avoid further contact with them at the earliest time possible. Student groups underrepresented in STEM occupations often perceive too few reasons to study STEM, and lack models that might stimulate their study. Many of these individuals underperform, thus limiting their future opportunities. External education supports are too few and usually available to only a limited number of students.

The MSS seeks to convert the strengths of STEM and its practitioners into promising and stimulating learning opportunities for **all** high school students. Critics are encouraged to identify faults with the model and make improvements.

Personnel Flow in the World of STEM

An analysis of the flow of human resources in the broad STEM world, as shown in Figure 2-2, suggests that few STEM personnel employed by private industry and government agencies traditionally contribute to high school STEM education, as the blue feedback lines are almost non-existent. With some notable exceptions, high school STEM educators stay within the high school and post-secondary education environments (represented by the shaded zone in Figure 2-2) and have limited contact with the non-academic STEM communities. The few companies, independent foundations, and government agencies that offer outstanding contributions to STEM education are a small fraction of an extensive potential resource, and impact a limited number of students. Usually, these resources require extraordinary efforts by individual teachers to be translated into classroom benefits, or they become minor appendages to the principal STEM instruction. The MSS attempts to activate the flow of STEM specialists (blue feedback lines) for catalytic effect while minimizing the need for additional work and responsibilities by teachers and changes in the post-secondary academic programs.

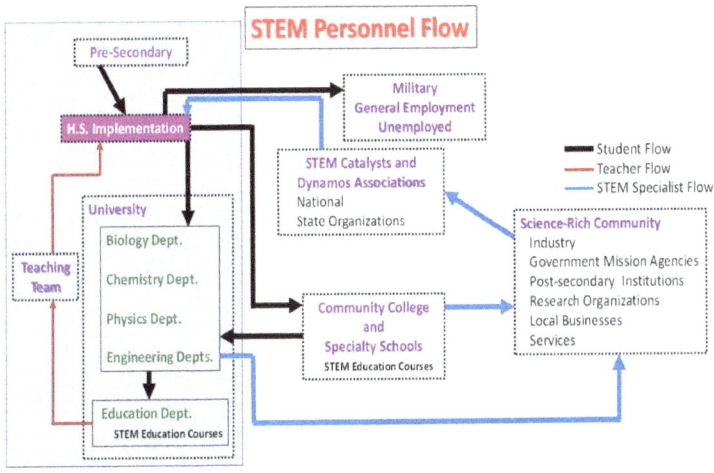

Figure 2-2: The Potential Personnel Relationships for MSS

Simultaneously, the capabilities of technology to support educational functions have been limited in traditional STEM courses to supplement instructional efforts episodically and satisfy school administrations' needs. Software for addressing complex data and varied inputs is standard in STEM operations and research but used sparingly to facilitate education and reduce teachers' workload. The MSS proposes that a major technological asset, a STEM Database Center (SDC), be developed to facilitate curriculum implementation, improve understanding of the status of individual students' knowledge and skills, and ease the burdens of teachers trying to enhance student connections with STEM resources. The SDC also includes and coordinates essential resources for teacher teams and schools that otherwise have access to minimal STEM assets, thus mitigating some inequities experienced by underrepresented minorities and schools of limited resources.

A bothersome perspective of some STEM teachers is that their sole function is to prepare students for the next step of education. The fact that most students will spend more than 80% of their next 50 years trying to put food on the table and keep a roof overhead seems foreign to some teachers who otherwise are deeply devoted to the current and future wellbeing of their students. Prepping students for the next possible class is essential, but using the unique characteristics of STEM courses to help students build their abilities to concentrate, communicate, analyze, calculate, cooperate, and participate in team projects provides life skills that may prove even more useful.

STEM teachers cannot be considered automatons that only teach the facts and procedures. Perspectives of students often change radically during their high school years. Hence, teachers always work with "moving targets" — students who need preparation for continuing academic work, entering the world of work, or adjusting to new life challenges. Simultaneously, the implementation of the facts of STEM content changes continually, thus requiring STEM teachers to make adjustments frequently to content and teaching strategies if they are to serve their students maximally.

Responding to future job markets is a risky matter if only specific skills and knowledge are considered necessary. Students should be encouraged to consider their preferences, the prognostications of experts, and all the data available to make wise choices about career preparation and college majors. By the late 1950s, information about plastics and computers was publicized widely in all media, but only attained limited coverage in academe. Within only a few years, applications of plastics ranged from consumer packaging and apparel to mechanical gears. Eventually, computer use became ubiquitous throughout the military, government agencies, and business. Scientists and engineers improved and applied these new materials and technologies, although they had never studied about them in college classrooms. They applied and adapted their knowledge, learning abilities, and problem-solving skills learned in earlier STEM studies.

Each class of students must be prepared through careful attention to broad foundational studies to address opportunities and conditions never considered in their pre-work classrooms. Success at enabling students to develop these broad STEM foundations may be diminished when students concentrate on currently popular specializations only. The students may later be impeded from entering other STEM careers, and the quality of the nation's workforce is reduced. However, specializations in attractive fields do provide strong motivational values for STEM study. Teachers must address these issues through curriculum building and providing students with sufficient time to study a range of topics.

The MSS described below is intended to harness national and local resources to improve student access to content, contexts, skills, and perspectives while reducing tedium for teachers. It provides opportunities to minimize inequities due to race, geography, and finances while enabling future employers to emphasize to students the basic requirements of employment and the characteristics of STEM practices used across the employment spectrum, from trades to cutting-edge research. It also provides teachers in classrooms the opportunities to address both broad content and specialties, encourage deep inquiry when appropriate, develop teamwork skills, profit from the experiences of many STEM professionals, and learn about a wide variety of career possibilities. It brings to students the human side of STEM and the impersonal but reliable nature of content and processes.

Parallel to the MSS, students should access career and technical education through programs focused on local resources. Becoming involved in a production or maintenance operation enables students to build confidence and develop skills and knowledge. Too often, vocational aspects of school programs and popular peripheral study are allowed to drive out more fundamental education, such as STEM courses. High school STEM education should be considered a foundational experience.

The Basic Theses of the MSS

A few of many possible comments suggest the depth of the problems compromising the preparation of students for 21st-century life and employment in STEM fields when using a 19th-century science education structure.

- High school science education provides the most severe constriction in the STEM workforce pipeline.
- The structure of high school science courses was established in the late 19th century when the promise of widespread assembly line manufacturing enamored educators. This structure reached the limit of its potential long ago.
- Teachers are overloaded with expectations and demands and are under-supported by the nonacademic STEM community.
- The potentials of technology applications in STEM education are underutilized.
- Both advanced and underdeveloped students are served poorly by much of current science instruction.
- Underrepresented student groups are usually unable to access role models, and often are enrolled in schools poorly prepared to teach many STEM topics.
- Many students leave STEM studies too early and have no familiarity with broad subject fields of STEM, although they need exposure to all STEM disciplines.

A Graphic Overview of the Model STEM System

Figure 2-3 shows the critical components of the MSS. Please interpret the figure as the situation a school would experience with the entire system operating at a steady-state. The brief component descriptions presented here are expanded in subsequent chapters.

Steady-State: A term used for a system to indicate that the start-up period has been completed and all variables have reached a condition of continuing stability. For example, a copier may reach steady-state conditions after a couple of minutes of warming up various components for operation; a new petroleum refinery may require several weeks for all its units to reach steady-state conditions.

Principal Components of the Model STEM System

1. **One STEM course** replaces the traditional 9th, 10th, and 11th-grade science courses of biology, chemistry, and physics. Students from all grade levels participate in the same class, providing each project team with one or more experienced members. This approach enables all students to participate in instruction that matches their knowledge/skill level rather than chronological age, and also facilitates peer teaching.

2. Students would engage in **large, complex projects** throughout the course. Students would manage and conduct each project under scrutiny and guidance by a lead teacher and a Catalyst STEM Specialist (CSS).

3. **Mini-courses** provide appropriate instruction to groups of students who are homogeneous in need and prerequisites (not by age or grade level). The STEM Database Center operations enable teachers to identify students expeditiously for homogeneous groupings.

4. A **teacher team** ideally consists of at least one teacher for each specialty of biology, chemistry, physics, and engineering. A regional or state STEM Service Center would supply Catalyst STEM Teachers (CST) upon request from a school to "complete" their existing teacher teams.

5. Regional or state **STEM Service Centers** (SSC) would provide: (1) CSTs to assist smaller schools and to fill specialty needs otherwise not available for any school; (2) consultation for developing

and changing school STEM curricula and activities; (3) professional training related to STEM instruction; and (4) limited facilities and services to support student projects generally along with materials otherwise not available in schools with exceptionally limited resources.

6. The **STEM Database Center (SDC)** would be a national organization (with branches to accommodate the volume of activities) that provides for: (1) access to education standards (national, state, and local) for curriculum design and testing/evaluation; (2) student status evaluations to facilitate assessing (diagnostic and formative) individual student's needs for specific instruction; (3) student project documents for downloading to local devices; (4) contacts for CSSs; (5) lists of instructional units and reviews; and (6) production facilities for developing and improving student projects.

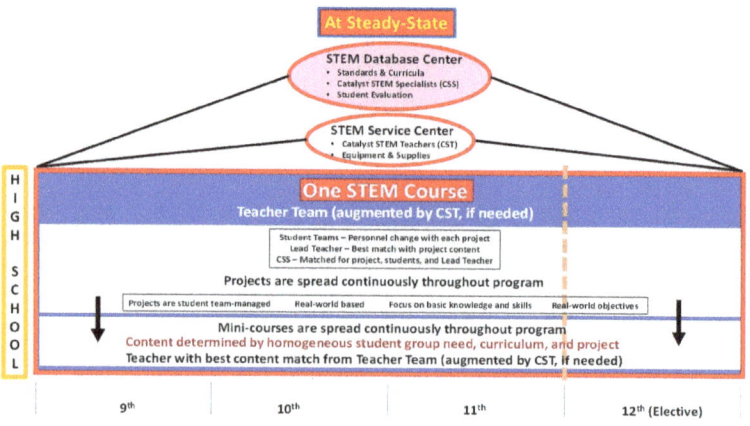

Figure 2-3: Diagram of All MSS Components

Conclusions

The MSS would facilitate individualized instruction, provide role models and career information, support teachers, enable students to overcome previous academic shortcomings and misconceptions, and develop skills valuable both inside and outside the workplace.

KENNETH M. CHAPMAN

3

One STEM Course

The Proposal: Replace the three traditional introductory science courses of biology, chemistry, and physics with a single, multi-year STEM course that can accommodate all sciences, engineering fields, and technology.

STEM is Multidisciplinary

Science, technology, engineering, and mathematics are the STEM constituents. The MSS assumes separate mathematics courses will continue to be available to all students. However, mathematics is vital to STEM and is often used to make sense of relationships in the S, T, and E components of STEM. A STEM teacher may need to provide mathematics instruction when required for a STEM topic. Many students enter high school STEM courses with severely limited skills in arithmetic as well as in algebra. STEM teachers may offer these students unique mini-courses early in the STEM course to provide sufficient arithmetic, algebra, and geometry proficiency for supporting study of specific topics.

Engineering may blend into science much more easily than will technology. However, many projects would offer substantial opportunities to address technological subjects both as tools to support science and engineering and as products for supporting and improving life. The MSS would provide blending opportunities for STEM topics usually taught only in specific disciplinary courses.

Course versus Courses

Most high school students have been offered science in three one-year, independent courses in biology, chemistry, and physics for more than a century. About 98% of high school students take a biology course. However, the comparable percentage drops to 76% for chemistry and 41% for physics.[19] Clearly, many American students receive a skewed view of science, and few become acquainted with engineering or technology. High school students frequently "drop out" of science instruction for a variety of reasons, including fear of mathematical content. Instead, these students often opt into courses of reduced scientific rigor that are accepted for state-science requirements, enroll in career and technical education programs, some of which lack rigorous STEM components, or are "excused" from more STEM coursework by sympathetic counselors or administrators. Many teenagers who avoid completing the traditional three-course series eliminate themselves from most opportunities for continued study in many programs that lead to work in STEM occupations and result in higher-paying careers. Also, these students reduce their abilities to make personal and civic decisions that involve STEM issues.

Presenting basic science in the siloed courses provides administrative advantages. Student transcripts are clear and understood universally. Recruiting teachers with expertise in a specific science field is relatively easy. Scheduling for these courses fits comfortably with the courses of other disciplines. Classrooms and laboratories can be equipped for single science disciplines, thus providing efficiency and reducing possible sources of tension or conflict. Each science teacher meets their classes in a room/laboratory that is controlled exclusively by the teacher. Instruction can take a path favored by the teacher without appraisal by other STEM-knowledgeable instructors. Some teachers may seek to change their course(s) continually to update content and teaching strategies and incorporate recent advancements in the field, while others repeat the same course from year to year. The silo arrangement reduces the complexities of teaching.

Specifications for the Single STEM Course

1. Addresses national, state, and local standards as required by the school's curriculum.

2. Is managed by a teacher team as a single, continuous course over three years with an elective fourth year.

3. Engages all students in student projects almost continuously and uses mini-courses to support and enhance student projects and address necessary STEM components not readily incorporated into high school student projects.

4. Applies all possible local, regional, and national resources to maximize student exposure to the practices of STEM and content knowledge and skills.

5. Requires a teacher team to maximize the instructional expertise available to each STEM student at all times.

6. Addresses all STEM disciplines applicable to projects so that students leaving the course at any point will have some familiarity with all the sciences, engineering, and technology.

7. Accommodates incoming students regardless of previous STEM experience.

8. Provides a comprehensive description of each student's progress in STEM to inform parents, guardians, administrators, schools to which a student may transfer, and other educational institutions.

9. Accommodates instructional practices best suited for each student.

10. Facilitates learning about and applying all disciplines that may contribute to addressing the objectives of any student project.

11. Enables the acquisition of knowledge and the development of skills for all significant components of STEM.

12. Improves both oral presentation and written communication skills.

13. Applies diagnostic and formative testing with analysis to ensure each student has appropriate foundation skills for understanding instruction necessary for mini-courses and student projects.

14. Uses homogeneous groupings for mini-courses to enable students to learn based on their present knowledge and skills of a topic rather than chronological age.

15. Enables student teams to be populated with students ranging from experienced to novices in STEM for specific projects.
16. Presents opportunities for students to develop and practice teamwork and leadership skills.
17. Provides access to career information for student development and decision-making.
18. Emphasizes inquiry as a learning method.
19. Emphasizes individual responsibility for learning.
20. Emphasizes the application of all of each individual's knowledge, skills, and experiences to problem-solving and decision-making.
21. Develops a sense of responsibility to other team members.
22. Emphasizes peer instruction as a means of learning and a skill that is useful in the workplace.
23. Results in two grades for each reporting period: (1) STEM knowledge and skills; and (2) understanding of STEM for future personal and civic responsibilities.

The Multi-year STEM Course

Replacing the three traditional science courses with a single STEM course in the MSS offers several advantages:

1. The course curriculum would offer a uniform opportunity for addressing all STEM disciplines. Content from basic to advanced topics would be taught when they are needed naturally for projects, rather than being constrained by a student's inadequate preparation or a teacher's inadequate knowledge about a specific topic. For example, some concepts taught only in a physics course may need an early introduction to make a project more effective. Likewise, instruction about some topics commonly taught in a 9th grade biology course might be delayed until the third year.
2. For student projects, team members would range from third-year (and fourth-year) students experienced with the course structure and expectations to first-year students unfamiliar with the course structure and having limited STEM knowledge.

3. Project management would rely on experienced students.
4. Students competent in specific STEM topics would serve as peer teachers.
5. Student projects would apply all disciplines as needed.
6. Members of the teacher team with the most expertise in a topic would be available to teach that topic whenever appropriate.
7. Students would not be required to revisit topics they already understand.
8. Students not meeting learning standards for a topic would be able to revisit the issue through several projects and mini-courses until the expected level of understanding is acquired (or they leave the course).
9. Students making rapid progress would advance to the extent they wish, and perhaps specialize in an area of interest. Some might complete the course requirements in less than the prescribed time, receive full course credit, and continue with more advanced work.

The course curriculum would address STEM content and processes. It would specify requirements for a variety of levels of completion as suggested in Chapter 9.

Students with Meager Qualifications

Some students enter the ninth grade well prepared to succeed in STEM work. Other students have had little contact with STEM activities or have not acquired the desired working skills in mathematics, measurement, or observations. Some students lack any interest in STEM subjects. Using a single course with mini-courses, projects, and extensive diagnostic and formative testing would enable students to be instructed individually or grouped homogeneously to address shortcomings and misconceptions rather than glossing over their inadequacies and continuing their failure mode. Later in the STEM program, some students with lackadaisical attitudes may change their perspectives and become ready to make rapid progress. Using some basic mini-courses to develop missing components in their foundation may enable many of these latter students to establish desired capabilities for future STEM work.

Students with Advanced Qualifications

Some students enter the ninth grade eager to tackle STEM topics and possess good mathematics and analytical qualifications. The one-course design, fortified with a well-constructed diagnostic and formative testing program, would enable these students to skip content already conquered and receive instruction at a challenging and stimulating level. These students may complete course requirements faster than prescribed and receive full credit for the complete course. They can then proceed with more advanced academic challenges. Some of these students may prefer to proceed in the class to achieve more advanced recognition, as suggested in Chapter 9.

Teachers

Working in a teacher team whose members have strengths in different disciplines would enable individual teachers to focus on their areas of expertise and avoid trying to teach content with which they have little familiarity. For example, some biology teachers consider themselves poorly qualified to address molecular structure issues in transport across membranes, while many chemistry teachers are fully qualified to do so. Other topics would find the teachers' competencies reversed. Many teachers would be delighted to challenge some students with advanced topics not usually included in the traditional setting. Content for which no teaching team member feels competent would be addressed appropriately by the Catalyst STEM Specialist (CSS) or a Catalyst STEM Teacher (CST) under the MSS.

Further, many education leaders expect science teachers to incorporate engineering and technology routinely into lessons and projects. The MSS mitigates this problem in two ways: (1) the projects are produced by teams that frequently should include practicing engineers or technologists, and (2) the teacher teams can incorporate CSSs who are engineers or technologists into many projects. A third, and preferred, way to improve engineering content is open to those schools that can employ an engineering teacher.

At steady-state, each STEM teaching team should require the SDC to assess their proposed project list for science, technology, and engineering relevance to promote equitable treatment of each disciple across the content of their school's curriculum. The same assessment should identify the available CSSs to provide balanced content treatment and career examples. The teacher team always retains the final judgment and can modify any assessment results to accommodate local needs.

If available, an engineering teacher should serve as the lead teacher for projects with a substantial engineering component. If an engineering teacher is unavailable, the lead teacher should request a CST become involved in order to address the engineering issues with the student team.

Production teams developing student projects should identify the engineering and technology components or options and address them thoroughly in the orientation video for lead teachers, CSSs, and CSTs. The project developers should always assume that some scientists may be CSSs for the project and may need extra preparation for the engineering components.

Curriculum

The teaching team would design the multi-year STEM course curriculum to address an appropriate portion of the standards applicable to their school. Content and examples would come from the variety of discipline strengths represented on the teaching team; however, projects with components outside the teaching team's expertise need not be avoided. Standards and additional content for which the teaching team may be unqualified would be addressed by a CSS and/or CST, thus giving students competent instruction in any subject desired for the curriculum or that might arise unexpectedly. The SDC should make available sample curricula to provide starting points for curriculum design by teaching teams. Teaching teams should incorporate their strengths and local resources into the creation of their school's curriculum. The school's curriculum would be the driving force for the diagnostic, formative, and summative tests developed and analyzed through the services provided by the SDC.

Teacher teams may wish to assign a few student projects frequently enough that some students may find themselves in teams tackling familiar problems. The design of projects would provide sufficient breadth that students with experience in a project would have new roles and leadership duties in a subsequent repetition of a project and have new content to be learned.

Administrators

A student's progress in the multi-year STEM course must be reported regularly and described by grades required by a school. A suggested grading/progress design is proposed in Chapter 9. Single summative grades commonly used provide a minimal view of each student's status in the course. Parents would have to be informed about the unusual course characteristics and the reporting of student status.

Students transferring into the STEM course would present few problems as the diagnostic/formative testing program provides an analysis by the SDC of each student's readiness to address the instruction appropriate at any point. New students would be introduced to their unfamiliar roles in the STEM system by students experienced with system operations.

Students transferring out of the multi-year course and into a traditional program present a much larger problem. The grading design suggested in Chapter 9 results in a thorough description of an individual student's progress. Still, it may need to be adjusted to describe progress on a typical traditional science course basis.

As applied to 9th-grade students, the STEM course curriculum may need to ensure a strong biology bias in assignments within projects and in the central theme of the prescribed mini-courses. Students leaving the STEM course would not have knowledge and skills precisely parallel to their counterparts in traditional courses during their first year, but they would possess some of the specific content and most of the skills. More advanced students would have been exposed to a significant portion of the traditional science courses and should not be significantly handicapped if they transfer into the traditional setting.

Publishers

The purveyors of all types of instructional materials should respond to the MSS without great difficulty. Many existing materials would need no changes. Other materials could be reorganized, and extant materials applied extensively. Discipline organizations could continue to focus on their areas of expertise and offer instructional units for student projects and mini-courses. They could contribute test items related to standards presented by the SDC. When SDC suggests teaching strategies to individual teachers, it also should identify relevant instructional materials from all organizations. Discipline organizations should be encouraged to have student projects of their design reviewed and included in SDC listings and descriptions.

Accreditation and Registries

Accrediting agencies would need to recognize the STEM course as equivalent to the three standard traditional science courses. Registries would continue to evaluate teachers in the disciplines and identify those qualified to work with the multi-year STEM course.

Conclusions

The single multi-year STEM course would promote individualized instruction, facilitate correcting student deficiencies, create opportunities for developing leadership skills, encourage peer instruction, and motivate some students to overcome their complacency or distaste of studying STEM. Each year of the course would have content from several disciplines spread throughout it, and discourage early exits from STEM instruction. Teaching team members would have continuous responsibilities throughout the course, teach topics for which they are qualified, and receive support for content for which they have limited qualifications. Students would receive evaluations at intervals dictated by the school, or more frequently.

References

19. National Science Board. Science & Engineering Indicators 2018 https://www.nsf.gov/statistics/2018/nsb20181/report/sections/ elementary-and-secondary-mathematics-and-science-education/high-school-coursetaking-in-mathematics-and-science (accessed Dec 21, 2021).

4

Student Projects

The Proposal: Comprehensive student projects would be one of the two key components of the teaching strategy for MSS.

General Perspectives: Student projects have long been a staple for teaching about STEM at all levels. Usually intended for a single student or a small team, teachers and a few organizations have designed projects of varied quality. (In addition, some organizations have enabled high school students to become participants in ongoing research efforts or to serve in internships.) Individual teachers with no external support direct almost all projects. Parents and larger audiences sometimes review student reports and project products. Panelists from outside the school sometimes judge student project results. Some projects incorporate career information and involve extensive media use for information, data development, and presentation. Projects may employ data collection from remote locations through government agencies and research organizations.

Student teams of up to 10 students would tackle each project in MSS. Teamwork may suggest to some readers the merging of contributions to give a single, shared result, thus lacking recognition of individual contributions. That is not the intent of MSS. Each team member should develop critical thinking habits and exercise creativity to prepare themselves for participating in team debates and negotiated compromises. Each student needs to build self-assurance that their ideas are worthwhile and influential, even when not adopted by the team. Students need to learn to:

- express themselves clearly and rationally rather than increasing their decibel level.
- extract clarification of knowledge and ideas from others.
- respect others as part of earning the respect of colleagues.
- understand that cooperation and compromise do not erase individuality.

Research on project-based learning has found that projects often have significant positive results in student grades and persistence in school attendance.[20] The characteristics of the projects suggested for the MSS include features not yet common for high school activities. Projects for MSS typically would be inquiry-driven and apply the four fundamental principles[21] necessary for effective project-based-learning: (1) construct from authentic settings and have a student motivating purpose; (2) integrate with other course content; (3) engage collaborative skills; and (4) use diagnostic and formative assessments to guide current and future teaching. The projects for MSS often would require more resources than do traditional counterparts. Although teacher-originated units would be encouraged, teams of experienced personnel having appropriate professional-level skills for their production roles would create most student projects for MSS.

Each student should engage in projects almost constantly in MSS. In addition to promoting learning about STEM knowledge and skills, each project would give the members of its student team opportunities to engage in: identifying objectives; developing strategic and tactical plans; assigning tasks and learning requirements for each team member; establishing and maintaining schedules and sequences; developing a report or product for both the lead teacher and the client (CSS); and acquiring career information.

Unlike traditional high school science experiments, the primary focus of a project would not be to have students follow prescribed, step-by-step instructions leading to specific expected answers. However, students frequently may need to use conventional experiments during projects to develop essential skills and practice measurements and observations.

Compared to many collegiate science projects, the projects for high school students would be much less open-ended than for more advanced education levels[22], more constrained to address content standards, and more time-restricted. Students would have access to much more expertise for each project – all teacher team members and a CSS – than in the traditional classroom.

Table 4-1 compares student projects for MSS with the traditional experiments used for high school science courses.

Table 4-1: Student Projects in MSS vs. Typical Science Course Experiments

Condition	Student Project in MSS	Typical Experiment
Team size	Up to 10 students of varied STEM experience	2 or 3 students of assumed equal skills
Time	2 weeks or longer	1 class period
Origin	Research, industry, or community; multiple STEM topics likely	Narrow focus for specific topic
Developer	Team of academic, STEM, and media specialists	Academic

Teamwork skill development	Major effort	None
Clients for project	CSS and lead teacher	Teacher
Student leadership	Required	None
Activity plan	Developed by student team with guidance	None
Objective(s)	Derived by student team	Given
Procedures	Planned by student team	Given
Learning assignments	Identified and assigned by student team	None
Work assignments	Identified and assigned by student team	Teacher
Evaluation of work	Student team, lead teacher, CSS	Teacher
Opportunity for improvement	Evaluation may require additional work	None
Communications	Oral and open-ended written draft reports to team and lead teacher and final reports required	Form completed or written final report
Learning of STEM content	Multi-discipline with variation by student	Single discipline and same for all students

Specifications for a High School Student STEM Project

1. Mimics the adult world by accommodating large student teams (up to 10 students) whose members have widely varying skill levels while enabling each member to make measurable contributions toward attaining the project objectives.
2. Addresses STEM problems and issues directly applicable to industry, research, or communities to provide opportunities for learning essential knowledge and skills that respond to education standards specified for local curricula.
3. Correlates with all applicable education standards and blends with a wide variety of typical high school MSS curricula.
4. Applies a generalized structure that enables the rapid development of new projects in response to motivational events in the world of STEM work.
5. Encourages each team member at their level of knowledge and skills in the project's subject matter to extend their understanding through self-directed learning, teacher intervention, peer teaching, and mini-courses.
6. Created by an adequately funded team that includes at least one STEM specialist in the project context, one or more high school STEM teachers, and a support team consisting of specialists in writing and media production.
7. Encourages team members to exercise creativity, critical thinking, and inquiry learning.
8. Requires planning and managing the project, communicating, and peer teaching responsibilities of the student team leaders.
9. Engages each team member in learning new information and skills, collecting data, communicating with all stakeholders, problem-solving, and reporting.
10. Accommodates students at all levels in advancing their learning significantly.
11. Applies a CSS as a client who also serves as a consultant for technical content and career information.
12. Ensures students have access to the maximum expertise possible through the lead teacher from the teaching team who is best qualified for the project contents, and facilitates access to other teacher team members or a CSS for additional assistance.
13. Enables students with limited access to career models to become aware of people from diverse backgrounds who have been successful in STEM work.
14. Enables teacher teams to employ a broad range of projects by accessing resources of CSSs who provide both expert instruction and materials not available otherwise.
15. Correlates with appropriate assessment tools and grading rubrics.
16. Accesses documents and resources provided through the SDC.

Reality in STEM Education: Hands-on science and engineering is a requirement of the MSS and includes planning, management, analysis, and interpretation activities as well as the manipulations that provide observations and data collection. Experiential activities both motivate engagement in minds-on work and create a sense of science and engineering practices that simulations and media cannot duplicate.

Hands-on student experiences create lasting impressions. Casual conversations and interviews with adults in non-STEM occupations frequently reveal that the most memorable parts of high school science courses are the experiments and demonstrations, even when the science content is not understood. Some STEM professionals describe substandard and misguided high school experiences in science courses that left them believing that practicing science or engineering would be a very different affair than what they had encountered in classes. Fortunately, the occasional fad of eliminating science experiments from high school science courses has been resisted successfully for many years.

Even demonstrations of science phenomena create memorable experiences. For example, a group of college students was overheard on a New Jersey beach enthusiastically discussing chemistry. Continued eavesdropping revealed the students were from Princeton University, where Professor Hubert Alyea, a world-renowned demonstrator of chemical reactions then nearing the end of his career, had created lasting impressions. At that time, many chemistry teachers were using Professor Alyea's safer demonstrations and his TOPS (Tested Overhead Projection Science) equipment to enliven and explain chemistry more effectively. One of Alyea's devotees, Professor Bassam Shakhashiri at the University of Wisconsin-Madison, regularly packed large lecture halls with his ticketed event, "Once Upon a Christmas Cheery, in the Lab of Shakhashiri," including many fans not pursuing STEM careers.

Often high school students have difficulty linking their STEM experiments with any external affairs or interests. The projects in MSS would emphasize the real contexts and practices of STEM and provide situations in which investigations have meaning and relationships with the world beyond the classroom.

The Curriculum and Projects: Building a curriculum with a strong emphasis on projects would significantly challenge each teacher team. The first guide should be the standards to be addressed during the multi-year course. Some standards would be addressed in many projects, while others may appear only once during the three-year course. Some team members would not work significantly on all the standards addressed by the activities within a given project. Balances among STEM disciplines, local resources, school facilities, student orientations, and teacher preferences would need to be set by the teacher team. Mini-courses would be used to bolster projects, mitigate student inadequacies, provide correlations among topics, and address standards deemed necessary but not covered sufficiently by anticipated project work. Each year, the teacher team would identify commitments of local resources before a curriculum could be considered complete. Any projects engaging unique local opportunities would need to be developed and prepared for implementation.

At the beginning of each school year, lead teachers should encourage student teams to select relatively uncomplicated projects to address small numbers of components requiring only simple mathematics. Novice students would learn about course features; advanced students would practice leadership skills in straightforward situations. With experience increasing during the year, students would tackle increasingly complex projects.

The teacher team would develop lists of projects so that each student would have opportunities to work with all the standards appropriate for their level of STEM expertise by the end of the school year. Thus, the lists would focus on sets of standards addressed and provide a steadily increasing complexity throughout the year.

Building the curriculum each year would be a challenging task for the teacher team. Construction of the curriculum would provide team members a clear picture of how the course should progress, their roles as lead teachers during the year, and requirements for supplies, equipment, and facilities. Upon completing the curriculum design, each team member should be skilled at interacting with the SDC and the technical aspects of using the testing program. The teacher team and school would determine the expectations of support from the SSC for the coming year and communicate that information to the SSC.

Contents of a Complete Project: A published complete project would include all appropriate materials for students, lead teachers, and CSSs. Most projects would require written, video, computer programs, and online components. Training materials for lead teachers and CSSs and classroom instructional materials would be part of each project package.

As a project production nears completion, the production team (and their supervisors, if any) would invite STEM specialists, experienced CSSs, STEM teachers, SCTs, and others to review and comment about content, instructional attributes, and production values. The final product would be a product vetted by representatives of all the stakeholder communities.

All of the materials would be available at or through the SDC, perhaps for a fee. Students would have direct access only to student materials.

Each project would be self-contained; thus, some redundancy with other projects would be expected. For novices, all the content would be new; for the experienced, some redundant content could be ignored, and some would provide a refresher. Some material would be new to all the team members. The lead teacher and project team leaders should give every team member challenges without creating overwhelming requirements.

Project Materials: Student materials would include an introduction using written and media materials that contain a description of the context within which the STEM content is applied and describe the relationships among the personnel at typical worksites. This introduction must include sufficient information to either imply or state the objective(s) directly so that the student team can determine what product or action will satisfactorily conclude the project. Student materials should include references to facilitate student research efforts for relevant information. The introduction to the students usually would be enriched by the CSS adding information unique to their familiar working environment and experience.

Components of a Project

1. Student materials: Introduction, context, sufficient information not available elsewhere to identify and address the objective, and references.

2. Lead teacher materials: Student materials, all CSS materials, suggestions for management, lists of questions, lists of skills needed, list of sources for necessary uncommon materials and devices, content not available readily, supportive media, a sample final report and suggested scoring rubric, suggested timelines, and references.

3. CSS materials: All student materials, all lead teacher materials, suggestions
    for an oral introduction with media support materials, lists of content that
    may help support the lead teacher, and suggestions for feedback to students on drafts of reports and the final team report.

4. Training materials for the lead teacher and the CSS: Background, STEM content, and suggestions for instruction.

5. Suggestions for handling anticipated problems and terminating the project.

Lead teachers would have continual access to general training materials for the MSS and the products available for individual projects (see below). In addition to the student materials, each project would provide lead teachers lists of questions that would lead eventually to specific assignments for individual students or groups of team members. A ranked list of skills should guide the students' development of specific techniques and observational qualities. For example, a project addressing acid/base chemistry might: Acquaint low-skill students with characteristics and applications of indicator papers; require more advanced students to perform titrations using strong acids and bases with common indicators and pH meters; and provide advanced students work with buffers.

Lead teachers also would be furnished sources for uncommon materials and devices that are likely to be needed. The unusual items also would be made available through the CSSs for donation or loan, depending upon the school's potential for financial support. The lead teacher would be provided a list of additional references at a more advanced level than supplied to students. Complete coverage of content not otherwise easily obtained would be provided. Lead teachers would be made aware of the video and computer-based media that students might use directly for self-study or through mini-courses. Lead teacher materials should include a suggested timeline for the project.

Project materials would provide the CSS suggestions for an oral introduction with supportive media and access to all student and lead teacher materials. Specific information about individual students would be provided to the CSS by the lead teacher (respecting privacy requirements) through the SDC. The CSS materials also would include suggestions for supporting the lead teacher and providing feedback on report drafts and the final report.

Self-contained Training Program for the CSS and Lead Teacher: The CSS would have had some experiences relevant to the core subject; the lead teacher may have no prior experience. An SSC would ensure one or more of its CSTs were skilled in the subject and ready to back up any implementation by a school. Specific projects could be the subject of training institutes, webinars, and other supportive activities.

Project Completion Materials: A sample student final report, suggested scoring rubrics, and suggestions about feedback to the student team would be provided for the lead teachers and CSS.

Project Termination: Project materials for the lead teacher and CSS should include information about indicators suggesting when a student project should be terminated. This material also should provide recommendations for maximizing student learning from such an event.

Production Characteristics Applicable to Each Project

Origin of Project:

Preferred: STEM organization (industry, non-profit, or community), which may be the financial sponsor when possible.

Satisfactory: Any organization or individual with a STEM story to tell.

Personnel:
Content co-directors/executive producers:
STEM specialist; high school teacher
Advisors: Other interested parties
Production team:
Professional media director
Education writer
Video producer
Computer programmer for instructional settings (Individuals may have multiple roles)
Reviewers for content and instructional values

Production site:
Conference room for 10 people
Art facilities
Video production equipment (studio and portable)
Computer facilities for production of all types of instruction

Registration:
To be listed with the SDC, a new project would be assessed for quality by reviewers under contract to the SDC.

Characteristics of Production Processes for Projects:

Origins: Traditional high school science experiments often have no context indicating how the content and skills to be learned have any applications beyond explaining the science. Teacher-originated projects sometimes give context, but seldom venture into the world where most STEM action occurs.

The broad spectrum of STEM activities in industry, non-profit organizations (including academic), and government agencies should provide the contextual subject matter for most projects in the MSS. Seldom would actual situations in their entirety be mimicked by a project. Instead, the production team would use the situational contexts to design projects that would challenge each student team member to develop specific skills and content knowledge required by a school's curriculum. For example, electrical circuits that control magnets in the CERN collider might provide novice team members opportunities to learn about the characteristics of simple series and parallel direct current circuits and how to measure electricity variables. More experienced students could explore alternating current circuits. Advanced students may study basic electronic control techniques. The CSS for project implementation may not be associated with CERN, but would have expert-level skills with the electrical circuits and equipment used in the collider. The lead teacher also might need support from a CST with much more knowledge about electrical circuits and measurement tools. Thus, students would learn from an instructional team to convey the excitement of a "big" physics effort searching for tiny particles. They also would learn about the characteristics of the people who work in environments like CERN, while also learning fundamentals about electricity useful for everyday life as well as in exotic science, engineering, and technology efforts.

By serving as producers of a project, content specialists would use their equipment and processes as examples where basic STEM knowledge and skills apply. Thus, a potential future employer would gain some name recognition and incorporate public information about some of the latest STEM applications into classrooms while contributing to improved STEM education. Sometimes, the organization seeking to add a new project or update an old one would become a financial sponsor of the production team effort. Connections between an originator and a production team often would take place through the SDC or an SSC.

Although the high school STEM learning needs may seem old and even trite, up-to-date contexts would make them exciting. A production team might update an outmoded project by changing only the context, even though what the students need to learn may not change.

Personnel: Production of the materials for a project would be a team affair. Designing the project and overall management would vest in the executive producers, a STEM specialist expert in the intended content, and an experienced STEM teacher. They may invite advisors to enrich the project further. They would provide initial outlines of the project, usually applying guidelines developed and modified during the creation of the initial projects for MSS. The production personnel may not be employed full-time as project creators; however, they should be provided sufficient time by their full-time employers to create the best possible projects.

SDC and other organizations would assemble support teams for project development composed of writers skilled in developing instructional materials and media specialists to interact with the executive producers. Each project should demonstrate highly professional production qualities.

Production Site: Production loci may be at the SDC, an SSC, a college/university, or other organization sites. Production equipment should include high-quality onsite and remote videotaping capabilities and computer graphics. Production personnel should be sufficiently expert in maximizing the instructional quality of both project materials and associated training aids. Funding would come from a sponsoring organization, STEM leadership organizations, or government agencies and private foundations. Production time for a project is likely to involve several months of elapsed time with several periods of multi-person production work. Both online and in-person meetings would be extensive. The production team would field-test the project in various settings before releasing it for general use.

The final production step for all purveyors of student projects would be to ensure the materials qualify for registration with the SDC and inclusion in the database for teacher review and application. Thus, teachers would have a single source for information and materials through the SDC. Control of training materials for CSSs and teachers would reside with the project owners.

The Production Process: At a steady-state, a catalog of hundreds of projects created by many different writing teams would be augmented continually by new additions, including some featuring recent headline events in science, engineering, and technology. Events like the discovery of the Higgs Boson in 2012 and the pandemic of COVID-19 in 2020 should stimulate the rapid development of new projects by experienced writing teams. New reviewed projects would start entering classrooms a few months after or even during the event, instead of being delayed for the extended periods required to get new textbooks into schools.

The SDC would provide digital connections to project materials and, when opportune, offer the actual materials through agreements with originators. Guidelines for production would encourage structural consistency for the projects, and mechanisms would be in place to provide peer reviews to ensure the high quality of both content and pedagogical elements.

Registration: Registration with the SDC would mark the completion of a project. Criteria for registration would be publicized and require periodic review by a quality assurance body. Immediately upon registration, the SDC would invite potential customers to review the project and the training materials. After an adequate number of CSSs reach acceptable readiness levels for working with student teams, teachers and schools would be informed about a project's availability.

Implementing a Project

In addition to the students, personnel for conducting a project include a lead teacher who manages the learning environment and a Catalyst STEM Specialist (CSS) who serves both as a client and a consultant. If necessary, a Catalyst STEM Teacher (CST) from the SSC serving the geographical area that includes the school may be needed to support the lead teacher or even become the lead teacher. The student team of up to 10 members of varied academic levels would provide one or two team leaders.

Implementation at a school would require acceptance of the project by the local STEM teacher team. Usually, a student team would be permitted to select from a list of projects chosen by the teacher team to address a specific portion of the school's STEM curriculum. Upon selection of a project, the lead teacher would use the SDC database to survey available CSSs qualified for servicing the specific project at no cost to the school (see Chapter 6). Negotiations would enable a CSS to be selected and scheduled. Project materials and teacher training materials would be accessed as needed. Work on the project would proceed as described below.

Each project would be a mutual learning event engaging the lead teacher and members of the student team. However, a suitable lead teacher may not be available at a school that is unable to employ a full complement of STEM teachers. In that circumstance, a CST from an SSC could support the school's lead teacher whose background is inadequate. In some instances, a CST may serve as the lead teacher. Thus, students in deprived schools could experience instruction equivalent to that expected at advantaged institutions.

Student teams would include up to 10 members from varied STEM education levels. Thus, most teams would consist of 9th, 10th, 11th, and perhaps some 12th-grade students. Mini-courses would underpin the project. Some students at advanced grade levels might not have achieved the knowledge and skills of STEM expected at their grade level. The lead teacher would assign each team member to mini-courses appropriate for skill levels determined by current assessment analysis programs at the beginning of the project. Thus, students would receive assignments and mini-course instruction at the proper level for their STEM qualifications.

One or two of the most advanced students would be identified as the team leaders and take responsible leadership roles under the lead teacher's tutelage and the CSS. They would lead team meetings, make assignments, follow up on assignments to judge progress toward meeting schedules and deadlines, and oversee the production of the final report. Their responsibilities would include identifying the instructional needs of team members and doing peer teaching when possible. Most of the interactions between the team and the CSS would flow through the team leaders and the lead teacher. The teaching team would ensure that every student would take team leadership positions for one or more projects before completing the STEM course.

Projects would require elapsed time periods as short as two weeks (e.g., identifying local insects) to several years (e.g., measuring trends in local insect populations). Individual students would engage in recommended instruction in mini-courses or self-directed study as indicated by the assessment program. Each teaching team member would be available to apply their specific expertise for mini-courses and consult with the lead teacher for a project. Personnel from an SSC would be available for consultation or as temporary members of a school's teaching team. All the physical resources for STEM education would be available through the school or the SSC to advance the work necessary for the project. Additional physical resources and services also might be provided by the CSS. Thus, students at impoverished schools would have all the essential resources to study STEM at appropriate levels.

When possible, a local context for the project would be used and reference real-world applications, manufacturing, research, environmental, ecological, or life issues. More complex projects would be scheduled later in the school year as the students improve their skills in STEM work and become more adept at teamwork and project management.

The Student Project Operation: Each project would have two foci: (1) promoting the learning of STEM content and processes and (2) developing teamwork skills.

Educational standards such as those presented by NGSS from the National Academies of Science, Engineering, and Medicine or specified by the state would provide extensive guidance for the specific content of each project. Lead teachers would manage their local instruction to achieve education standards goals addressed by the project AND by applying their awareness of their students' characteristics, the assessment program, and personal preferences. Standards found not to be addressed through projects but required by the school's curriculum would be subjects of required mini-courses. Mini-courses would be offered as appropriate to match individual student needs and capabilities. Some students might need to address knowledge usually developed at earlier grade levels. Students desiring to exceed the curriculum expectations could access mini-courses at advanced levels or through guided self-study. Thus, all members of the student team would advance their knowledge and skills for STEM.

Robert Bly wrote an article for practicing chemical engineers in which he noted that, "It does not matter whether you are a people person, happy to work shoulder-to-shoulder to tackle a project, or an introvert, at home in the lab or behind a computer. The corporate culture of today demands that you work well with others, regardless of your preference."[23] The skill of working well with others starts developing very early in life. With many students not completing education programs beyond high school, developing teamwork skills throughout the high school years should be enhanced through projects, in addition to learning about STEM content and practices.

There are several interpersonal skills STEM students need for their future, irrespective of their activities following high school. The table that follows presents some of these skills (adapted from Bly's article).

The student team should develop a plan for the work of the project. The team may be encouraged to use the lead teacher's favorite approach to planning and managing a project, or use a procedure suggested in Appendix B. After being provided an introduction to the project by the CSS, the team would ensure each member understands the objective(s) and what the product(s) should be -- a report, a crystalline chemical product, a building structure, an electronic circuit that meets specifications, a set of architectural drawings, a model, etc.

Interpersonal Skills Everyone Needs to Develop

1. Respect others by learning their names quickly and frequently using each individual's name preference in conversation and when referenced.
2. Listen more and talk less to maximize learning. An essential skill is to ask thoughtful questions to stimulate a conversation or to drive discussion in the necessary direction to reach decision points.
3. Remember that unsolicited advice is likely worth its cost of zero. However, contribute comments about safety and identify erroneous information tactfully and timely whenever relevant.
4. Recognize that promoting oneself is seldom a desirable activity. Actions and results will bolster recognition.
5. Avoid adopting a superior attitude.
6. Develop good manners. "Please," "thank you," and a friendly attitude helps to develop cooperation.
7. Avoid cursing. Using curse words suggests a limited vocabulary.
8. Follow Benjamin Franklin's dictum, "Speak ill of no one, but speak all the good you know of everybody.*"
9. Control tone and demeanor, even when agitated. Going into attack mode usually engenders resistance to anything one may wish to achieve.
10. Exercise good hygiene and health practices in family life, the workplace, and in all social gatherings.
11. Recognize you may be one who likes working alone and solving technical problems. However, conversing and working with others always becomes necessary, whether one is a plumber instructing a customer about what not to flush down a commode, or a chemist who needs to persuade others that a new Nobel prize-worthy pharmaceutical compound has a market.

*https://www.goodreads.com/quotes/81547-speak-ill-of-no-man-but-speak-all-the-good (accessed Dec 21, 2021).

Each project would provide a variety of activities for challenging assignments for each student at their skill level. Novice STEM students would be assigned basic activities in which they gain basic STEM knowledge and skills. Advanced students should lead by applying their current STEM expertise to solve one or more problems and also acquire any additional abilities they would need or desire. Practice to gain proficiency would be required frequently, and typical traditional science experiments may help develop proficiency. For example, novices may use pH paper to judge acidity or basicity and gain familiarity with the pH scale. More advanced students might use pH meters to compare reality with theoretical pH calculations. Advanced students may measure pH with data-logging devices during titration of weak bases with strong acids and interpret the curves of plotted data.

Each project would contain quantitative elements to ensure all students engage directly in mathematical work or appreciate the importance of quantitative measures and calculations. Imbedded in project activities would be questions that encourage each student to exercise their creativity.

Conclusions

Student projects are one of the two principal instructional components of MSS, the other being mini-courses. In addition to learning STEM content, student projects would teach skills for planning projects, managing activities, collecting data, analyzing information, teamwork, and communicating. The projects bring CSSs into direct contact with students to serve as clients and models, and also provide career information. Applying the resources of the SSCs and CSSs would enable schools with non-ideal resources to offer their students far more opportunities than with traditional science classes.

References

20. Baines, A., et al. Key Principles for Project-Based Learning; Lucas Education Research: San Rafael, CA, 2021 (accessed Dec 21, 2021).

21. *Ibid*

22. Cianfrani, C.; Hews, S., and DeJong, C. Student-Driven Research in the First Year: Building Science Skills and Creating Community. Journal of College Science Teaching, Vol. 50, Issue 2, 58-68.

23. Bly, R.W. Interpersonal Skills for Chemical Engineers. *Chemical Engineering Progress*, March 2020.

KENNETH M. CHAPMAN

5

Mini-Courses

The Proposal: Single-concept mini-courses for homogeneous student groups would be one of the two key components of the teaching strategy for MSS. Mini-courses also include self-study for individuals or small groups.

Mini-courses complement projects as the other component of instruction in MSS. Mini-courses are short, single-purpose units of instruction lasting from a few minutes to a few days for homogenous student groups having similar readiness and need for a subject. Mini-courses would lack some of the careful sequencing and spiraling of the traditional courses. However, many mini-courses in the MSS would benefit significantly from the immediacy of application in a current project. Each mini-course would have a narrow objective presented as a single concept, or as few new concepts and ideas as possible.

The teacher team would assemble students with similar readiness and need for a specific mini-course using data analysis by the SDC of diagnostic and formative tests from individual students and the teacher team's knowledge about student performance. Usually, the teacher team would not admit a student lacking adequate prerequisites to a mini-course.

Although most mini-courses would address a single concept, some would correlate several topics or provide broad overviews to show relationships among multiple topics and disciplines. Some students might use only media presentations. Homework and practice would follow.

Instruction would be provided by the most appropriate teacher team member or by a CST if relevant expertise was unavailable at a school.

Specifications for MSS Mini-courses

1. Enable curricular requirements (based on recognized standards, state dictates, and local needs) to be implemented during the three-year course.
2. Support student projects.
3. Address theoretical concepts, practical STEM operations and measurements, or tactile activities and experiments.
4. May be repeated by students who wish to reach higher levels of mastery.
5. Concentrate on a single discipline or may be multidisciplinary. (Students in Year 1 may engage in a preponderance of biology content.)
6. May require students to leave project work to participate in mini-courses.
7. Provide maximum flexibility for responding to curriculum requirements as well as to student needs and readiness to address learning objectives.
8. Address any topic that contributes to STEM understanding and processes.
9. Address immediate needs of a group (one or more) of students requiring the content for the curriculum, a project, or other activity.
10. May be used to correlate topics and give overviews and orientations.
11. Duration is determined by the need and may range from a few minutes to several class periods.
12. Apply to students of similar readiness for the content, irrespective of experience in the course (readiness is determined by the SDC analysis of student data and teacher decisions).
13. Are administered by the teaching team member or members most effective for the content or by a CST.
14. May be presented by media.
15. Require student action with homework, tests, experiments, etc.
16. Provide grade results at two levels: (1) for STEM competence; and (2) for citizen/life applicability. Only the highest grades earned for specific content would be part of a student's permanent record and calculated GPA.

The regular traditional class assumes all students have familiarity or mastery of the prerequisites for the immediate subject and that each student has the same interest level. Students inadequately prepared fall further behind; advanced students are bored and see their presence as a waste of time or irrelevant.

The MSS uses the SDC to help teacher teams group students homogeneously for each mini-course. Students lacking the prerequisite competency may encounter the content of a scheduled mini-course later when they have addressed their inadequacies satisfactorily. An advanced student may need the instruction from a mini-course populated by novices due to missing the content previously or failure to reach the desired level of performance for the subject. The teacher team may advise lower-level students who have satisfied the prerequisites to take a mini-course for advanced students.

The curriculum adopted by the school would include the application of the standards to be met by students, which in turn would drive the assembly of tests and analysis of test results. The curriculum would specify required mini-courses and preferred sequencing. The teacher team would add mini-courses to meet student needs, such as correcting deficiencies in readiness to benefit from instruction in the MSS. The SDC operation may recommend specific mini-courses to the teacher team based upon its analysis of student tests matched to the curriculum and the projects selected. Mini-courses would be taught or supervised by the teacher best qualified (and available) in the teacher team. The "assigned" teacher would be responsible for all aspects of the mini-course and oversight of students using media in place of class sessions.

Some students may need to participate in a specific mini-course more than once to reach an acceptable level of mastery. Mastery, in this case, is a judgment in which both student and teacher would have roles. A student adamantly avoiding a future job involving STEM may be satisfied with a barely passing grade for the subject using the "citizen" grade and then failing the assessment for STEM knowledge and skills. Later, the same student may decide a STEM job actually *is* in their future, and retake the mini-course to improve their mastery level. Students who enter the MSS with inadequate preparation or are unmotivated may become eligible to take introductory mini-courses after many months in the MSS. They would still have the opportunity to catch up with classmates or make substantial progress toward developing STEM career capabilities.

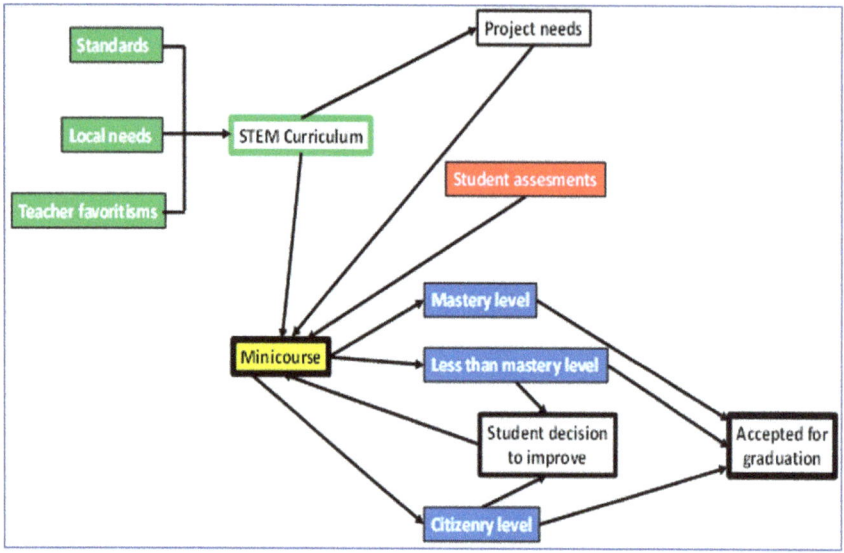

Figure 5-1: The inputs and outputs for each mini-course

The rare student with outstanding capabilities and interest in STEM could do independent study in MSS, with the close oversight of the teacher team and regular participation in the student projects. Some of these students might prefer doing some projects independently. They would do these projects in cooperation with a CSS and possibly a CST, with oversight from the teacher team. The mini-course arrangement would facilitate providing advanced instruction without compromising teaching for the base curriculum.

Assessments

Most mini-courses would use three types of assessments: (1) diagnostic to determine the readiness of the student to participate in the study; (2) formative to determine progress; and (3) summative to determine the degree to which students have learned the objectives.

Assessments would start with test items from SDC suggestions for the standards to be addressed, with teacher modifications as desired. The teaching team would transmit raw results to SDC to analyze. SDC would return a comprehensive report of test results and suggest the next steps of instruction.

Table 5-1: Sample Mini-courses Involving pH Concepts

Level	Content
Basic (1 hour)	Describes pH and Basic pH Measurements
Intermediate (2 hours)	Relates pH to Acid and Base Concentration; Basic Calculations
Advanced (3 or more hours)	pH and Equilibrium; Calculations

SDC would be a repository of test items for diagnostic needs for common mini-courses. It also would have a similar bank of test items to evaluate the degree to which students have achieved the desired understanding of each standard. Teachers would evaluate each test item suggested by SDC for their own testing needs and make additions as they deem appropriate. SDC would have analysis capabilities to give objective results and recommend next steps and materials to support those steps. SDC resources would enable diagnostic assessments reaching back to elementary arithmetic and perspectives about STEM.

Students would be encouraged to understand that most testing is to facilitate learning, rather than being punitive. Only results of final tests would become part of the permanent record of the student.

Support from Publishers and Discipline Organizations

Almost every school would use many of the same mini-courses. Companies in the publishing community would develop a broad spectrum of print and other media materials to facilitate mini-courses. A variety of technologies could deliver the media to facilitate both classroom and individual student use. SDC would be a convenient source of information to teachers about all available support materials, including reviews. Teachers would not need to waste time searching for materials to meet their needs, nor overlook materials that would be very helpful.

Discipline organizations should not need to change current procedures significantly to advocate for and support specific fields. Their materials that meet SDC requirements would be listed within the database and brought to teachers' attention as appropriate. They may find developing specific mini-courses to be advantageous.

Conclusions

Mini-courses offer broad flexibility for addressing the needs and desires of individual students. Many students enter ninth grade with deficiencies that ensure either their poor performance in MSS or limit the opportunities of the class meeting desired objectives. Mini-courses offer a technique for redressing the needs of individual students without causing better-prepared classmates to be bored with repetition. Combined with a grading approach suggested later, mini-courses enable students to raise themselves to as advanced a level as they desire. By the end of three years, the instructional strategy using mini-courses would ensure students' command of STEM content knowledge and skills should equal or be superior to students who have taken the traditional courses.

6

The Teaching and
Learning Corps

The Proposal: Teaching teams at each school would design and implement a curriculum and would be augmented to ensure students would have access to a high level of expertise for STEM content needed to support projects and other learning endeavors. Students would be taught at the level of their expertise in STEM knowledge and skill.

The Teaching Corps: MSS engages a variety of personnel for teaching STEM: (1) a teacher team would be composed of the teachers typically employed by a school; (2) for each student project, a Catalyst STEM Specialist (CSS) not otherwise connected to the school would serve as both a client and mentor for the student team and lead teacher; and (3) many schools having less than ideal teaching resources would use one or more Catalyst STEM Teachers (CST) from the local SSC. Thus, students at schools with limited teaching resources would have access to the necessary expertise to receive instruction equal to that of their colleagues at large and well-supported schools.

The Learning Corps: MSS uses a single course spread over three years, and may include a fourth year as an elective. Thus, each STEM class would consist of students from 9th through 11th grades, and possibly also seniors. MSS classes meeting at a specific time would include many more students than traditional courses. However, the teacher team would divide the class into homogeneous groups for mini-courses and student teams for projects of no more than ten students each. The entire teacher team would be available to any student for assistance and assignments during each mini-course and project. Thus, the teacher team would plan and organize before each class period to ensure the full engagement of each student.

Advanced students would engage in peer teaching as part of their project team leadership responsibilities. Some students likely would be involved in media-delivered self-study units and assessments. Overall, the STEM classroom should be very busy with students doing many different activities, just as their interests, needs, and levels of STEM knowledge vary greatly. Teachers would depend mainly upon the analytic and database services of the SDC to organize homogenous student groups and direct individual student activities.

Teachers also would be members of the learning corps through their work with CSSs and perhaps with CSTs. These professional working relationships would provide almost continual expansion of knowledge and skills related to STEM implementations in the "real world."

The School's Employees

The Teacher Team: High school teachers of STEM subjects are remarkable and dedicated persons who must satisfy many different roles. They must be:

- Scientists, engineers, and technicians.
- Aware of current advances in their fields.
- Adept at helping others understand technical content and the practices of science, engineering, and technology.
- Aware of the societal implications of STEM applications and processes.
- Sensitive to the issues of maturation of their students.
- Advisors about the future opportunities for students making critical decisions about life and career that will impact the rest of their lives.

Helping teachers fill these roles successfully is a crucial objective of MSS.

In MSS, teachers would be members of a team of up to four individuals, and continue to control learning efforts as in the traditional course arrangement. However, they would use a different teaching structure; share several key responsibilities traditionally dependent on a single individual; and apply more resources. Individual mini-courses and projects usually would be managed and taught by one teacher, with planning being a teacher team responsibility and project engagement being a student team operation.

The teacher team and its members would control their school's inputs and outputs from the SDC, the SSC, the CSSs, and CSTs. Persons and organizations involved in traditional high school STEM extracurricular education would continue to do so, perhaps with a few changes. Competitions, internships, and camps would continue to help prepare students for future roles in STEM. Some features of the MSS would improve support for the many students who are unable to participate in extracurricular activities.

Teacher Teams at Steady-state: For the MSS, the ideal teaching team would consist of four members with B.S. or higher degrees in three different scientific fields, and one would be an engineer. In addition to meeting the general requirements for teaching positions in traditional high schools, each member also would have extra preparation for directing teamwork and interfacing with the SDC and an SSC. Since only a tiny fraction of schools would be able to employ such teams, CSTs would be used to help achieve most of the results expected from an ideal team. Thus, the students from small schools would not be disadvantaged by lack of access to school employees who are experts in specific STEM content. The CSS would also compensate for any lack of technical expertise at the school needed in any particular project.

Specifications for STEM Teachers in MSS

1. Earned baccalaureate degree (minimum) in a STEM discipline* or formal recognition as possessing high-level expertise in a STEM field of work.
2. Dedicated to maximizing the learning of each student.
3. Experienced or completed student teaching for a STEM discipline.
4. Experienced with non-academic work in a STEM discipline.
5. Trained for work as a member of a teacher team in an MSS.
6. Trained to direct student project teams in MSS.
7. Committed to regular updating of their technical content and teaching expertise.

* "Discipline" includes trades and technical service work as well as science, engineering, and technology.

At the first implementation of the MSS, the teacher team(s) would develop a curriculum. The team might adopt a suggested curriculum from the SDC, modify an existing curriculum, or create a new one that may add standards unique to the school. The curriculum would be adjusted as desired by the teaching team. The curriculum would be linked to the SDC services that make diagnostic and formative student tests, provide student assessments, suggest instructional activities, and connect to projects, resources, and CSSs. At steady-state, the school's curriculum should be stable. Through the SDC, the teachers would have available the tools to create diagnostic and formative tests, assess students, identify homogeneous student groupings for mini-courses, suggest instructional strategies to address individual student needs, provide projects, and identify and provide communications to CSSs. Thus, teachers would be relieved of several typical duties and have more time for interacting with and teaching students.

Teachers at Steady-state: A high school implementing the MSS would likely need at least two full years from initially enrolling students in the system to reach a steady-state. By that point, teachers would have experience with all components of the MSS and would have available students with two or more years of experience to serve as leaders and managers for projects. The school's STEM curriculum, stabilized to meet local needs and preferences, would guide the efforts of the teacher team. Teachers using the services of an SSC would be familiar with the available resources and CSTs.

Teachers would expect the SDC to provide convenient access to all appropriate national and state standards and to have the capability of adding many more standards statements from individual schools. Teachers would request from the SDC diagnostic, formative, and summative test suggestions that support the local curriculum. They also would be able to add their unique questions and answers to their school's files in the SDC. Using these resources, the SDC would assess student test results for teachers and suggest the next steps for instructing individual students. Teachers also would use the SDC to identify and link to CSSs and teaching resources. Figure 6-1 compares resources that teachers would have available for teaching between the traditional course arrangement and MSS.

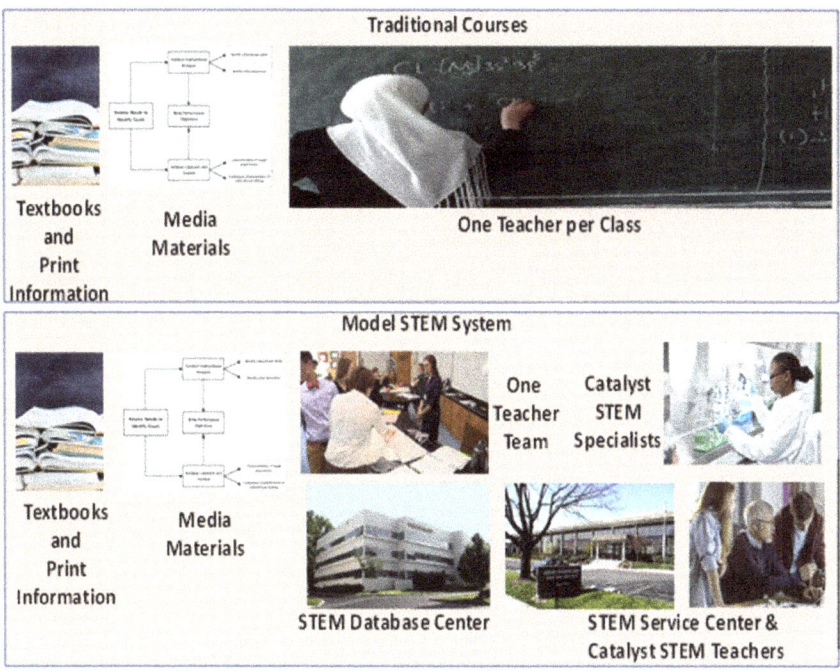

Figure 6-1: Resources available for teaching a traditional science program compared to those provided in an MSS.

Students

Learners at Steady State: The "average" student, or learner, would begin MSS as a 9th-grader. However, students may enter the MSS at any time, since assessment instruments would identify inadequacies a student might have for using a specific mini-course or project effectively. Teachers working with MSS could address any remediation promptly. Students leaving MSS before their third year might encounter a challenging situation when entering traditional science courses. For example, students leaving MSS would not have "completed" the equivalent of a single conventional science course until the end of the third year. However, most students leaving the STEM course at any time would have some familiarity with all science and engineering fields. The school would furnish student leavers with a comprehensive statement of their progress referenced to the standards they would have addressed.

Students new to MSS would enter a flexible instructional environment that accommodates students with widely varied readiness to study any specific STEM topic. Entering students, whether 9th graders or transfers, first would encounter diagnostic tests from the SDC to identify misconceptions and inadequate background for the particular school's STEM curriculum. Regardless of prior experience, students would be grouped homogeneously by their instructional needs for mini-courses, or teachers would guide them to individual self-instruction media. New students would be placed in project teams as soon as possible. Lead teachers would recommend to CSSs for special attention students achieving high levels of expertise. Figure 6-2 is a simplified comparison of student opportunities between the MSS and the traditional course structure. Note that students may exit further STEM instruction with the classic courses having had little exposure to physics or chemistry. That would be unlikely with MSS.

Students who perform poorly in their early participation in MSS and later become interested in preparing for a STEM career may start remediating their background at any time to overcome deficiencies. With teacher guidance, such students may repeat mini-courses, use self-instruction media, or benefit from peer instruction. Students may leave ongoing project activities as necessary to engage in mini-courses and other instructional activities.

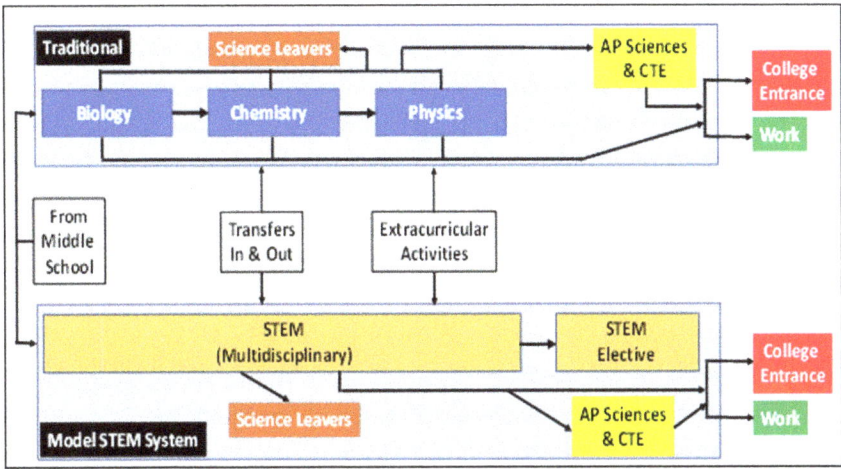

Figure 6-2: Contrast between student opportunities in traditional programs and MSS.

Specifications for Students Completing the MSS Curriculum

Students prepared for continuing preparation for a career in a STEM field will have:

1. Completed three or more years in an MSS curriculum.
2. Achieved at an A/B level for continuing in a science or engineering discipline.
3. Performed at a passing level for continuing any advanced work in a STEM field or CTE.
4. Demonstrated competence as an effective and efficient team member.
5. Learned to formulate questions to extract information necessary for effective decision-making and develop plans for the next steps.
6. Demonstrated competence in preparing reports for both team mission advancement and the team's final report.
7. Learned to develop a learning plan for oneself.
8. An understanding of the role and use of STEM as a citizen.

Students prepared for citizenry will have:

1. Exposure to a wide variety of STEM disciplines.
2. Differentiated between science, engineering, and technological missions and procedures.
3. Awareness of the importance of data, its accuracy, and its analysis for decision-making in any circumstance.
4. Competence in using the techniques of STEM for information collection and decision-making.
5. Ability to formulate questions to extract information necessary for effective decision-making and developing plans for the next steps.
6. Ability to develop practical plans to achieve a specified mission.

Through projects, students would be encouraged = or required = to accept assignments that call for learning new knowledge and skills. Advanced students serve as peer teachers for the benefit of both the learners and the peer teachers. Lead teachers would guide instruction and do direct teaching as necessary. Projects would involve students in planning, project management, evaluation, reporting, and providing opportunities for learning the concepts and practices of science, engineering, and technology. As their skills improve, students would be encouraged to take leadership positions in teams.

Students always would have available the best quality teaching resources the school could provide, plus the external resources of CSSs and CSTs. CSSs would extend the content competence of advanced students and provide career information to all students.

Catalyst STEM Specialist (CSS)

CSSs: A Catalyst STEM Specialist (CSS) would be an individual regularly working as a STEM specialist or perhaps a post-secondary student preparing for a STEM occupational area. Scientists, engineers, and technicians working in industry, government, or technical research organizations are potential candidates for this new opportunity. Thus, HVAC technicians, chemical process technicians, biochemists, nurses, internists, epidemiologists, chemists, electrical engineers, computer programmers, surveyors, veterinary assistants, carpenters, astronomers, astronauts, automotive mechanics, farmers, and other specialists who require expertise in one or more STEM fields may qualify as CSSs. For some projects, collegiate personnel may be ideal. However, most CSSs used for a project in a course should be non-academics to make students familiar with practitioners in a wide variety of STEM operations. A well-balanced curriculum would require projects that give students access to CSSs from a wide variety of backgrounds.

These representatives of the broader STEM community are a new addition to the high school education scene. Each project would have a CSS as a client/consultant. The CSS usually would work remotely through media connections to eliminate time lost for travel. Each CSS would train to interact effectively with high school students and become thoroughly knowledgeable about the content of the projects for which she claims competence. The SDC and other providers would offer this training primarily through media. Eventually, students preparing for STEM careers in college could take a course, perhaps for a single credit, that would provide initial preparation for roles as CSSs.

Specifications for CSSs

1. An employed or retired high-level practitioner in research, design, STEM operations, STEM trade, or STEM service area, or advanced student in a STEM specialty.
2. Trained to work effectively with high school students and teachers through student projects.
3. Trained to support specific projects.
4. Enthusiastic about and committed to improving student outcomes in STEM.
5. Able to commit to specific time appointments for interactions with students using virtual communications techniques.
6. Sustained by the employer or school.
7. Meets requirements for registering with the SDC.

CSSs would be obligated to become trained to interact effectively with high school students through projects. Then, they must become qualified for specific projects to be listed in the SDC. The lead teacher and CSS would negotiate a mutual understanding of the extent of the CSS's involvement in a project. Each project should require no more than an aggregated total of 8 to 10 hours total of the CSS's time. The CSS would spread his time for a specific project throughout the activity, perhaps over two-to-four or more weeks. Each CSS would determine the number of assignments they would undertake in a year.

Most CSSs would remain at their worksite(s) and interact with students remotely unless personnel near a school would be involved and agree to visit the school.

Recruitment: Some individuals may self-identify as CSS candidates as soon as they learn about the opportunity. Recruitment activities would reach most potential CSSs through their employers. Organizations that do not place a high value on public service and volunteerism would provide few CSSs. STEM organizations having a pre-college education mission, government agencies, and educators would try to persuade the CEOs of organizations that employ STEM specialists to identify some of their employees to serve as CSSs. CEOs must ensure employees receive benefits from acting as CSSs for student projects.

CSS Engagement: CSSs would deliver an introduction to the project to the students; review the plan proposed by the student team; receive oral status reports at mutually acceptable intervals; coordinate with the teacher about technical issues; and provide instruction for technical content particularly for students with advanced capabilities. They also would seek opportunities to discuss career paths and current work, if permitted by the employer. Many employers would encourage their employees to become CSSs for enlightened self-interest. Others may need state and federal government encouragement through tax advantages, incentives, and regulations.

Credentials: To be listed in the SDC, CSSs would demonstrate at least a minimum level of expertise to interact effectively with high school students engaged in projects. Candidates interested in becoming CSSs would receive training at their convenience, probably using non-classroom mechanisms. Video programs and webinars may prove effective delivery systems. Eventually, the B.S. graduates of STEM programs would complete a curriculum that requires a one-credit course providing initial preparation for service as a CSS. Periodic refresher training would be required to maintain credentials.

In addition to the initial preparation, CSSs need some training for each project they expect to support. (A CSS training video would be part of the production products for each student project.)

Stepping through CSS Involvement in a Project: Engagement of a CSS in a project might reflect the following pattern:

1. The lead teacher would review the list of potential CSSs for a specific project. The list would provide information about each candidate and their availability.
2. The lead teacher would contact the preferred candidate through SDC online forms and provide information about the school, the student team, and herself.
3. The candidate would indicate whether or not they would accept the invitation.
4. When the teacher receives an acceptance, a contract establishing the services to be provided by the CSS would also identify dates, time availability and limitations, and other details. Sample contracts from SDC would be available to facilitate this step.
5. The CSS would be introduced to the student team at the first contact and then present an introduction to the project.
6. With the lead teacher controlling all interactions between the students and the CSS, the team would present their preliminary plan for review.
7. After a student plan is accepted, the CSS would engage with the students, participate as a client and content expert in team activities, and make safety recommendations as appropriate.
8. The lead teacher and CSS would consult regularly and mutually determine if a project is failing and cannot be saved.

9. A project concludes with a report by the students, which must meet the expectations of both the client (CSS) and the lead teacher.

10. The CSS would submit a final grade for the project to the lead teacher, with detailed comments about the project and the report. The lead teacher would not be obligated to accept the CSS's grade suggestion.

Training and Preparation: CSSs need not be developed as teachers but must have appropriate training to interact effectively and efficiently with students, lead teachers, and the instructional environment.

A series of training programs or equivalent classroom work would provide general preparation for a CSS. Media programs would be made available universally at no charge. An assessment program would ensure a CSS candidate achieved a satisfactory level of expertise for engaging in an MSS.

A CSS would need to complete training for each project they propose to serve. The training materials would be developed by the production team for each student project and updated as necessary. The CSS would need to complete an assessment satisfactorily for each project. Some CSSs would serve only one student project; some would handle many projects. Each CSS and their employer would determine the extent of a CSS's involvement.

Catalyst STEM Teacher (CST)

Ensuring Student Access to Expertise in Content: Many schools would need the services of one or more CSTs to fill gaps in the expertise of a teacher team at a specific school. Small schools may have too few STEM teachers to provide four-member teacher teams. Available teacher candidates for a teacher team and previous circumstances may prevent having each major STEM discipline represented in a teacher team even in a large, well-supported school. Engineering teachers may prove challenging to hire for many schools.

CSTs should be available remotely to help STEM programs provide expertise to students in fields not represented adequately by the school's STEM faculty or for topics requiring particular proficiency. The extent of service required of a CST would vary from advising for a specific project to providing much instruction for a discipline area, such as physics. The CST would seek to satisfy the intent of the curriculum but would not be a regular member of the teaching team nor undertake additional teacher duties such as hall monitors, committee assignments, etc. CSTs would be made available from an SSC through a cost schedule that partially supports the resource center, reflects the school's ability to pay, and discourages schools from not hiring needed STEM teachers.

Recruitment of CSTs: To be employed by an SSC, a CST would be required to have directly relevant classroom teaching experience, preferably in a school using MSS. The CST would have demonstrated outstanding performance in team operations, earned advanced degrees in a STEM discipline, accumulated applicable academic preparation in education, and validated teaching expertise. With selection based on merit, not seniority, most would be hired directly from high schools at attractive salaries. A CST position would command high prestige in the STEM teaching community.

Credentials for CSTs: CSTs would have earned all the certifications required by the state in which they work. Also, they would be required to obtain additional national credentials as appropriate. For example, engineering teachers would seek recognition as Professional Engineers through the testing (and experience) program of the state in which they serve. CSTs would demonstrate a high level of professionalism through memberships in appropriate local, state, and national STEM education organizations, serve on committees and task forces, and some may serve as officers.

Specifications for CSTs

1. Experienced in classroom applications of an MSS.
2. Prepared for the content of one or more STEM fields at an M.S. or higher level.
3. Completed a high school teaching credential program or have equivalent experience.
4. Taught in a high school STEM field full-time for a minimum of five years.
5. Worked in a non-academic STEM setting for two or more years.
6. Taught in an MSS-type setting for one or more years.
7. Interfaces effectively with non-academic STEM personnel at all levels.
8. Effective as a teacher in both online and classroom settings.
9. Effective as a guide to students making career and college selection decisions.
10. Effective as a user and trainer of the SDC.
11. Able to travel.
12. Communicates well using virtual techniques.
13. Demonstrates effective team leadership skills in an educational setting.
14. Capable of working with several learning arrangements simultaneously.
15. Effective as a teacher and coach of teachers.
16. Updates her STEM background through participating in on-site STEM work in one or more STEM fields in industry or research settings and/or continuing STEM education in a post-secondary environment.

Training and Preparation for CSTs: New CSTs already would have accumulated substantial expertise in teaching STEM to high school students. However, they likely would need immediate training for the additional responsibilities for their new working assignments. Mentoring the new CST would be very important for the first year. A new CST with no recent relevant college classes or non-academic work would undertake refresher and updating work in their STEM discipline as soon as possible. Introductory training for CSTs would sensitize them to their new working relationships with teachers, students, and CSSs. Also, all CSTs frequently would seek to extend their expertise in both educational and STEM employment needs.

Conclusions

High school STEM students in an MSS would have access to a high level and quality of content and teaching expertise in all the disciplines required for the curriculum. The MSS teacher corps at a school would consist of the teacher team members employed by the institution, CSTs as needed, and supplemented by numerous CSSs for the student projects. Thus, students in small or inadequately-resourced schools would receive similar preparation for continued study or the workplace as their contemporaries in large schools with substantial resources.

KENNETH M. CHAPMAN

7

STEM Service Centers

The Proposal: Create and operate a centralized multifunctional computer database and programs to facilitate high school STEM teaching.

Enabling Disadvantaged Students: American public and private high schools range from large, well-funded institutions with well-supported teachers in all STEM disciplines to schools with small enrollments or limited funding. Even when enrollments justify four or more STEM teachers, a school may be unable to attract an appropriately diverse cadre of STEM teachers, or to provide adequate support for laboratory and external experiences for all students. Many schools also often serve disadvantaged students who lack access to role models in STEM occupations and encouragement to perform well and enroll in more STEM classes. These conditions deny many students adequate preparation for competitive post-secondary STEM education and opportunities in STEM careers. Such loss is unfortunate for individuals, employers, and the general society. STEM Service Centers could mitigate many educational inequities in high school STEM education through their STEM Catalyst Teachers (SCTs), professional development, and the loan or donation of facilities, materials, and technical services.

Components of an SSC: STEM Service Centers would serve either a state or a large population area by providing resources for STEM information and facilities for many disciplines directly to high schools.

Each SSC would offer a substantial amount and variety of supplies, equipment, and services to schools as loans or at a reasonable cost.

Each SSC would have a director to manage human and physical resources, promote STEM education, and demonstrate outstanding team leadership. The SSC staff would address the other missions of the center.

A vital function of the CSS would be to hire and support CSTs. Each center would employ teachers with high-level technical expertise in at least one STEM discipline (M.S. degree level) and experience in teaching high school students. The cadre of CSTs at each SSC would offer the schools they serve expertise for classroom support in various STEM disciplines not limited to biology, chemistry, and physics as offered in traditional science programs. In addition, the CSTs would deliver professional training for STEM teachers to individual schools, conferences, and workshops. They also would provide advice and assistance to schools for developing curricula and managing the local MSS, including interfacing with local employers.

Services of an SSC: STEM Service Centers would offer three primary benefits for all high schools: (1) a cadre of well-prepared STEM Catalyst Teachers; (2) loans or donations of supplies and equipment needed for experiential learning and (3) professional development for teachers and consultation on curriculum design and other STEM education needs. CSTs would be available from SSCs to provide professional development for STEM teachers, consult with schools and their teachers about curricula development and teaching, and fill teaching gaps at schools, either remotely or in person.

The comprehensive array of services provided by SSCs would alleviate many of the problems that significantly reduce the benefits all students should receive from their high school STEM program. Students attending schools serving impoverished communities and having limited fund-raising abilities could access improved opportunities to prepare for STEM careers or acquire information and talents needed to assume civic duties.

Each SSC would supply materials and equipment to schools that do not have and cannot afford the necessary resources to implement fully desired projects. It also would loan non-consumables as necessary. In addition, the SSC would identify institutions willing to support the unique needs of specific projects. The Center also would provide facilities for unusual measurements, production of artifacts, and identifications commonly unavailable in schools but needed to support projects. For example, synthesized chemical products from a project might need spectral analysis, or strength testing might be required for heat-treated materials.

Specifications for Each SSC

1. Overseen by a board of directors composed of parents with children enrolled in a high school STEM program, active STEM teachers, school administrators, and local employers of STEM personnel with representation across the spectrum of STEM employment.
2. Employs a staff including a manager, support staff, and CSTs.
3. Organized to maximize the opportunities of all students for learning experiences in STEM in all the high schools (public and private) in a specified region or state.
4. Serves homeschoolers cooperating in arrangements that concentrate students to work in teams.
5. Provides facilities to maximize the effect of virtual classes.
6. Offers professional development to all teachers.
7. Maintains an inventory and distribution system of supplies and equipment not usually found in high schools to support most student projects used in its service area.
8. Employs a cadre of teachers having expertise in many different STEM disciplines for service as CSTs.
9. Cultivates effective communications and cooperation between employers and the high school STEM education efforts in its service area.
10. Maintains effective two-way communications between the SDC and the high school STEM educators.
11. Supports CSTs participation in professional activities both within the region of service, the state, and nationally.
12. Supports the development of new student projects.
13. Manages the relationships between CSTs and the schools served.
14. May add missions unique to the area served.

SSCs also may become sites for creating new student projects, with CSTs serving on the production teams. Such work would balance the workload of the CSTs.

SSCs would develop the expertise of their staff members to interact effectively and efficiently with organizations providing CSSs. SSC personnel would sponsor

professional development efforts and assist in teacher training efforts by colleges and other organizations. They would offer reviewers to organizations supplying instructional aids and equipment for high school STEM instruction.

SSCs would coordinate work experiences and internships for students and teachers in organizations conducting scientific research, engineering design and implementation, materials production, and other services. (The experiences of workplace learning seldom can be provided in most academic institutions.) Establishing beneficial relationships leading to these experiences is time-consuming and requires long-term consistency. Each SSC would work on behalf of the schools it serves to coordinate students, teachers, and organizations offering extracurricular opportunities such as cooperative education jobs, internships, and competitions. Schools establishing cooperative education and internship opportunities in non-academic settings for high school students must address many legal and safety hurdles. However, the necessary efforts can lead to rewarding workplace experiences for students and teachers. SSCs would distribute the opportunities among the schools and individuals seeking to participate.

SSCs would advise and assist the SDC to identify problems and seek solutions for high school STEM education. They would be critical intermediaries between schools and policy-makers at state and national levels.

Each SSC would provide a two-way contact point for each school in its service area to identify problems that inhibit good STEM education and disseminate information about resources and new initiatives. The

SSC would ensure that high-quality STEM education becomes accessible to all students regardless of location and financial circumstances.

Support for an SCC: Each center would be supported partially through state funding and solicited donations and grants. Center personnel would be encouraged to advance their expertise by participating, for pay, in all the types of work in which STEM specialists work. Each SSC would be funded partially by fees paid by the schools receiving services. The SSC and each school would negotiate fee schedules based on specific income potential from a tax base or parental income.

Conclusions

The brevity of this chapter may demonstrate an inverse relationship to the importance of SSCs. The mission of each SSC would be to raise the level of STEM education available to each high school student to that enjoyed by students in the best schools. While costly, the SSCs would substantially increase the number of students well prepared for employment in STEM workplaces and ready for challenging continued education in STEM. The increased size of the taxpaying STEM workforce and reduced societal costs of failed students would offset the costs.

KENNETH M. CHAPMAN

8

STEM Database Center (SDC)

The Proposal: Create and operate a centralized multifunctional computer database and programs to facilitate high school STEM teaching.

A National Resource: Teachers use computers in varied ways to improve instructional processes and information acquisition and processing. They use computers for professional development from sources that would otherwise be unsuitable based on time or geography. During school closures for a pandemic, many students' only access to teachers and learning processes was through computer-based communications. Administrators and teachers use computers to build and share student records. Teachers often prepare daily progress statements for each student to provide almost continuous reporting to parents and others. However, using an SDC operation that is separate and independent of any governmental or specific organization would:

- Enable teachers to apply the guidance of standards continuously to implement their curriculum.
- Facilitate a diagnostic and formative testing program to identify flaws in individual student readiness for the proposed instruction.
- Help teachers create and implement instruction tailored to individual student needs.

- Identify instructional resources and facilitate communications with personnel applicable to individual student and project needs.

Additionally, teachers would gain more time for preparation and actual instruction.

Developing and operating an SDC would exceed the capabilities of a teacher, a school, a school system, and some states. The SDC would require a considerable investment in preparative work and programming, followed by continuing operations, maintenance, and user education. To handle the high volume of teacher contacts and the systematic analysis of tests from more than 12 million students (via teachers only) throughout the year and provide redundancy, several independently operating computer sites would likely be necessary. However, the potential gain in student results would make this a sound national investment.

To better protect student confidentiality, accommodate local control of unique or sensitive information, and relieve traffic focused on a few powerful computers, some components of the SDC might be made available for individual school use on local computers. However, the SDC must have a centralized operation to facilitate information updating and communications. User representatives, primarily teachers, must guide the development of the details for each SDC function described here.

SDC Governance

The principal goal of the SDC is to enhance teachers' effectiveness in teaching STEM. However, it also has the potential of having dictatorial power that ultimately could compromise the effectiveness of STEM education. Thus, the governance of SDC must be a significant issue, and it must be designed to give stakeholders – classroom teachers, parents, and employers – complete control of the organization, perhaps through a representative body called a Council.

Specifications for the STEM Database Center

1. Oversight provided by a Board of Directors composed of: (1) three computer database experts; (2) three classroom STEM teachers; (3) three leaders of STEM discipline membership organizations; (4) three employers of STEM personnel; and (5) three SSC directors.
2. Operated by a full-time manager and staff representing teachers as well as computer experts.
3. Presents easily used and robust user interfaces designed for interactions by STEM teachers and curriculum developers of limited computer proficiency.
4. Provides robust security for all users and data providers. No student information will be stored other than for an immediate transaction (e.g., for test analysis and building instructional suggestions).
5. Uses common listing characteristics to present STEM standards statements for both national and state-approved compilations. Teachers must be able to add locally-derived standards statements.
6. Enables non-SDC staff to add sample curricula with permission from SDC.
7. Presents sample curricula and allows modifications with permission from SDC.
8. Provides downloadable sample curricula for local review and printing.
9. Provides a test creation system that accesses and selects suggested test items at the direction of teachers from a repository of evaluated test items. Each test item must be adjustable by the requesting teacher.
10. Provides a system to analyze individual student test results for both free-form and objective responses.
11. Presents information about instructional resources that is updated constantly with monitored reviews.
12. Offers teachers suggested instructional alternatives for students having specific immediate learning needs as identified by an SDC evaluation.
13. Maintains a current list of suggested instructional materials for teacher consideration, including evaluated student projects, textbooks, media offerings, and materials to facilitate laboratory observations and data collection.
14. Creates and maintains lists and communication links for CSSs certified as clients and mentors for specific student projects. Augmentation with teacher comments must be possible.

One of the research projects studying the efficacy of MSS should create a draft document to describe the SDC Council's membership, governance responsibilities, standing committees, and operations. Suggestions for consideration include:

1. Council membership should be restricted and evenly divided between in-service STEM teachers and the other three principal stakeholder groups for high school STEM education – employers, parents, and post-secondary representatives.
2. The Council would have officers, including a president serving a one-year term. It would manage a succession plan to ensure continuity, experience, and training of potential officers.
3. Council members should be subject to term limits, with staggered terms to ensure a mixture of both experienced and new members would always be engaged.
4. The Council would identify its committees and prescribe responsibilities for each one.
5. Standing and *ad hoc* committee chair appointments or elections should rotate evenly across the four stakeholder groups.
6. Each operational unit of the SDC staff would be overseen by and report to a standing Council committee.
7. Council members would elect a Board of Directors from the Council's membership. Board members would provide each of the stakeholder groups equal representation and have staggered terms. Board members would serve terms of double the length of Council terms. The Board would have fiduciary responsibilities for the SDC, oversight of policy implementation, and direct the administrators of staff personnel.
8. Governing documents should provide for the removal of members of both the Council and the Board.

The Inputs and Outputs of the SDC

The user-friendly, multipurpose national database operation would facilitate:

1. Presenting national, state, and local standards.
2. Providing convenient uploading of locally derived standards.
3. Enabling individual schools, school systems, and districts to design curricula to fit local objectives and opportunities.
4. Providing approved student projects.
5. Offering formative and summative test items for local use.
6. Analyzing test results and individual student records to evaluate student status and progress.
7. Identifying students for homogeneous groups to be taught in mini-courses and by other means.
8. Suggesting instructional aids for individual students.
9. Listing CSSs and providing linkages for connecting a school with appropriate CSSs.
10. Suggesting guidelines and specifications for developing student projects.

Figure 8-1 suggests the inputs required and outputs expected of the SDC. Since the SDC would engage at least weekly with every STEM teacher during the school year, a very robust and reliable computer system would be necessary. Regional branches of the SDC likely would be needed to cope with the immense volume of interactions and processing efforts due to data inputs and analyses involving some 12 million students. Most, perhaps all, processing of student data likely would be handled locally to reduce SDC computer requirements and improve security. However, database updating would be a continuous operation. Further, the maintenance of a current database of instructional materials would require a national perspective. The SDC would also develop student project materials continuously.

An Interface Design Task Force

A crucial first step in creating and improving the database would be developing a user-friendly interface that meets all the criteria set by an Interface Design Task Force of STEM teachers. Members of the task force would include STEM teachers whose characteristics span the full spectrum of:

1. Age (entry-level to near retirement).
2. Teaching experience (none to 40+ years).
3. Computer literacy (novice to competent programmer).
4. Role (classroom teacher to national policy-maker).

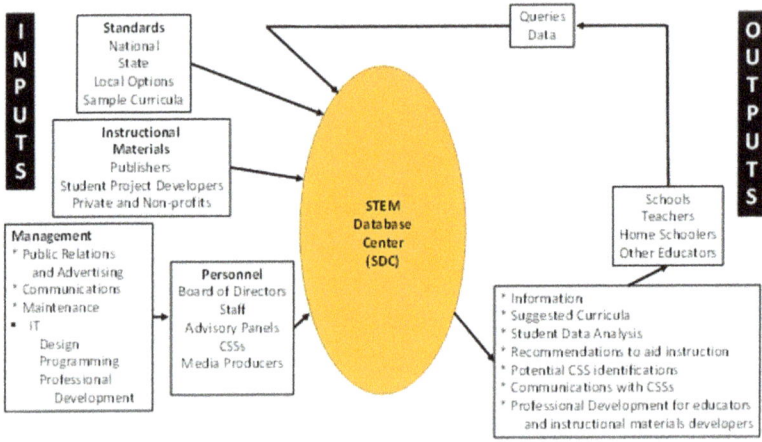

Figure 8-1: The Inputs and Outputs of the STEM Database Center

The task force management would ensure each member has the opportunity to express their advocacy and reservations on each topic. No single member or clique would be allowed to monopolize either discussion or decision-making. The working drafts of the interface submitted to the programmers must satisfy the needs of the teachers most challenged by the specific work to be facilitated by each interface.

As the database develops, the task force would review the efficacies of the draft interfaces and request changes as necessary. The database designers and programmers must comprehensively address each task force member's reservations and questions throughout the development process.

The task force would consider one principal customer: an overburdened semi-computer literate STEM teacher nearing retirement. This customer must encounter an interface that is user-friendly, intuitive, and requires no training. To achieve this end, the Interface Design Task Force must initiate the interface development prior to any work on coding. A draft interface that meets the desires of the task force would be accepted before any other work begins on computer program development. The Interface Design Task Force would interact frequently with the SDC programming unit from initiation to completion of the initial database tools and for all revisions thereafter. The Interface Design Task Force would start with only a few members but should add several teachers each month during the principal database development effort to ensure teachers encountering the database for the first time can use it effectively and efficiently.

Presenting Standards in a Database: Standards such as those developed for the National Science Education Standards (NSES) and Next Generation Science Standards (NGSS) and by many states would be the cornerstone for the database. Together with teacher additions, they would be the references for test items and student data analyses. The database developers may need to rephrase some of the original standards statements to help teachers design curricula and convert the standards effectively to classroom lessons. The same presentation also would be helpful to the creators of new instructional and reference materials.

The standards may need to be presented at multiple levels:

- Level 1: A unique and straightforward statement of each standard, intended as a reminder for teachers and content developers who are already familiar with the standard's substance and context as expressed for Level 3.

- Level 2: A statement that describes the specific intent and character of the standard. Perhaps the standard originator would write the statement tailored to the standard.
- Level 3: Each Level 2 statement would be expanded by describing contexts for using and judging the standard. This description may include correlations and linkages to other standards.

Combined, the levels for a given standard would present a comprehensive picture of the standard as intended by the developers. Accompanying each standard in the database would be a set of many test questions to apply to each type of student, as described in Chapter 9. Representatives of the developers of the standards should review and approve all database presentations of their work and the test questions.

Searching the Standards: User groups should design the interface for searching the standards in the database to ensure all users can achieve desired and useable results.

Curriculum Interfacing: A teacher team at an individual school should be able to interface with the standards database to construct a curriculum specific to their school. A teacher team should have various sample curricula furnished by the SDC from which to choose a potential starting point for developing a curriculum unique for their school. Standards originators, state or district representatives, and disciplinary organizations should create sample curricula to describe their perspectives about the content of STEM instruction. To produce their final curriculum, teacher teams should modify the sample curricula to eliminate or add standards and add locally-derived standards and comments. Each teacher team should add their adopted curriculum to the SDC database to facilitate modifications and notifications of any changes in standards and accommodate other applications as described below.

Diagnostic and Formative Testing: Identifying a student's readiness or need for specific instruction is often determined by a teacher's assumptions. Identifying misconceptions held by a student often relies on random encounters between students and teachers. Post-secondary teachers of recent high school students soon discover the frequent necessity of ridding students of misconceptions that handicap learning. What a high school student is taught sometimes is not what they learn.

For each standard, the SDC would offer a set of formative test questions appropriate for students at various levels of progress in STEM. Project developers would use the questions as guides in their creation of materials and suggestions for instruction. The SDC may suggest test items for a teacher's use; however, teachers would control the test items used. The responses to these questions should be interfaceable with SDC so student profiles can be developed to help teachers determine how best to proceed with instruction for each student. By interfacing with information about a school's class, the database would facilitate grouping students by their command of a subject so a teacher could develop a mini-course or an instructional program for a group of students homogeneous in their knowledge of a specific topic.

Teacher aids: Teachers would find the SDC helpful as it provides:

- Searchable standards statements.
- Sample curricula for adoption with modifications to meet local needs and desires or comparison to a locally-derived curriculum.
- Suggested diagnostic and formative test items selectable by criteria set by individual teachers and their curricula.
- Student test results analysis, which the database would then use to develop recommendations for the instruction of individuals and specifically identified homogenous groups of students.
- Suggested instructional resources.
- Materials for student projects for downloading.
- Lists of CSSs for projects and communications links to facilitate developing agreements for services.
- Teacher training on the use of the center's resources.

- Relevant notifications for teachers and other interested parties as appropriate.

Chapters 6 and 8 suggest the database requirements for addressing issues related to the CSSs.

Support: The SDC would operate as a non-profit organization. The federal government would provide financial support through grants from agencies. Foundations, service fees based on schools' ability to pay; and employers would provide additional support.

Employees: The SDC would be an essential independent cog in the STEM education system for the nation. It would need to cooperate with STEM teachers, professional organizations of STEM personnel, employers, education personnel from states and locales, post-secondary institutions, government agencies, and private and non-profit supporters of STEM education. Thus, the SDC would need a strong but cooperative administration focused on helping to provide the best STEM education possible for all American high school students. The administrators would include educators and STEM professionals from across the full workforce spectrum, from trades to advanced research. Positions would have tenure limits to provide a steady inflow of new ideas and perceptions for an educational arena in constant flux. Personnel should include many "rotators" (teachers, other school officials, education leaders, and graduate students employed for short terms as SDC employees).

In addition to the continual development and maintenance of the complex database operations, management of student project reviews, and the stimulation of new student projects would be ongoing responsibilities of SDC. This effort would require both experienced teachers and a continually renewed cadre of STEM professionals. A possible non-exclusive commitment would be creating new student projects requiring a permanent media production staff.

Conclusions

The SDC would be a site of leadership and innovation in STEM education with no enforcement power for the nation. It would require support from many sources. It would significantly reduce the constant tedium of support activities that engage teachers before and after their classroom work, while also contributing to the instruction of individual students. The SDC would be a resource of primary importance for fostering excellence in STEM education for all schools and each student. It would provide resources for all schools and all students to promote the strongest possible STEM education.

KENNETH M. CHAPMAN

9

Assessing Student Status and Progress

> **Proposal:** Create a bifurcated grading system to distinguish between STEM content knowledge and STEM for citizenry; Create a ranking system to display level of STEM knowledge relative to expectations implied by standards.

Designing appropriate questions for tests in STEM and judging student responses is a mixture of science and art. For example, analysis and discussions sometimes require many meetings spread over several years before the Examinations Committee of the American Chemical Society accepts a new question for its standardized exams. For MSS, initial development and maintenance of a question inventory at SDC may be the most challenging activity.

To augment the standards statements, the SDC would maintain examinations committees to develop extensive lists of questions for each standard for use by teachers (and others) as described in more detail below. The SDC also would provide convenient interfaces and apply artificial intelligence to enable correlation with a school's curriculum for making teachers' selection of test questions fast and easy. SDC would provide analysis of test results through suitable interfaces. Retention of student data would be severely restricted, if permitted at all.

STEM instruction addresses two main goals in MSS: (1) preparation for additional STEM study (resulting in a STEM Mastery grade) and (2) improvement of life skills and preparation for citizenship (resulting in a STEM citizenship grade). Providing a bifurcated grading/reporting system for students would offer several advantages: (1) STEM content need not be "watered down" to allow more students to achieve passing grades for the course; (2) all students would recognize STEM instruction provides essential life skills, some of which originate uniquely in STEM; and (3) all students gain an understanding that the goals and practices of STEM are meaningful to civic decision-making. Individual schools or districts might drop the STEM Mastery grade from GPA calculations to encourage students to continue with MSS. However, students would be encouraged continually to succeed with all elements of STEM content. Those who had not achieved early success in the STEM Mastery category would be able at any time to begin doing so, due to the flexibility and testing program of MSS.

Assessment of student performance on projects would have subjective components to add to opportunities for objective knowledge and skill improvement evaluation. With contributions from the CSS for the project, lead teachers would provide subjective judgments, perhaps following the student project developers' suggestions.

The following discussion assumes that teachers would: have total control for the selection and acceptance of questions from SDC; provide the final assessment for all types of tests used at their school; and be able to use their own assessment questions and instruments. SDC devices would enable teachers to insert their unique questions to assess specific knowledge and skills that fit their school's curriculum and instructional program. Other administrative levels – district, state, and national – also would find the extensive SDC question banks helpful. Also assumed is that teachers sometimes would admit students not meeting readiness measures to higher-level instruction with the expectation that the students would overcome known deficits.

Many instructors present their introductory college STEM courses expecting that students will not bring a uniform set of knowledge and skills from previous instruction to the class. Most topics start with a short treatment of introductory material that should be a review/ reminder for many students, and unknown material for others. If they have prepared adequately, students of the latter type would persevere and soon reach par with their classmates. Otherwise, they spend extra time in college or drop out of further study of STEM content. MSS test analysis would provide teachers with detailed information about each student's status in STEM. MSS provides detailed reports based on specified STEM education standards and a school's curriculum to demonstrate each student's level, progress rate, and goals relative to further STEM study. Parents and administrators may be satisfied with summary letter grades for tests and end of marking periods – or they can obtain detailed status reports.

Specifications for the SDC Assessment Component*

1. SDC supports committees and staff to monitor assessment devices, remove obsolete test questions, and create new questions.
2. Test questions designed for appropriate student levels would accompany each standard statement for diagnostic, formative, and summative assessments.
3. Each student project must include questions at levels appropriate for diagnostic, formative, and summative assessments.
4. Using a school's curriculum as a guide, SDC must create tests from its databases as requested by a teacher.
5. Teachers may add their own questions and assignments with answers and scoring rubrics to tests created through the SDC.
6. SDC programs are accessible by teachers to grade student tests.
7. SDC programs analyze each student's test to identify deficiencies that may hamper performance and strengths that may suggest faster progress.
8. Using class and student performance information, SDC programs will suggest homogeneous student groups for mini-courses and list appropriate instructional materials to aid teachers.

*Much more detailed specifications should be developed by task forces of teachers prior to SDC database program development.

The STEM course for MSS lacks the typical time and sequence characteristics of the traditional courses. Students entering as new 9th-graders would be considered 1st-year students and receive grades at semester end for STEM 1. Students transferring from other high schools would enter the STEM course at the 10th, 11th, or 12th grade and receive their first-semester end grade for STEM 2, STEM 3, or STEM 4. Thus, a student changing schools after the beginning of the second semester of the sophomore year and being enrolled previously in chemistry would receive their first grade in MSS for STEM 2, second semester.

Students lacking prior participation in a program like MSS would require preliminary preparation and diagnostic testing similar to that provided to new 9th-graders. A 10th-grade student transferring out of the MSS course and into a chemistry course would not have studied some of the content familiar to the new classmates, but would use skills developed in MSS to catch up quickly.

The STEM course for the MSS is oriented more toward mastery of knowledge and skills of the content and procedures addressed than to a chronological exposure to knowledge and skills for a specific set of topics. Most testing would make extensive use of SDC resources and analysis capabilities.

MSS uses four types of assessments:

Diagnostic – Used to help the teacher team learn about each student's background and readiness for addressing STEM instruction for specific content in the MSS. This testing would reveal exceptional shortfalls of knowledge and skills that must be addressed before a student can derive substantial benefit from the STEM instruction in a mini-course or project.

Formative – Used frequently to guide individual and small-group instruction by disclosing each student's progress for specific topics in a project or mini-course.

Benchmark – Broad summary tests that help identify students ready for increased rank in MSS, as described below.

Summative – Used for determining required reporting grades. Although necessary for administrative and counseling purposes, these grades have limited use for directing instructional efforts compared to diagnostic and formative tests.

Diagnostic and Formative Tests and Testing

Sets of questions at different levels would accompany each standard and each student project presented in the database of the SDC. These questions would evaluate a student's: (1) readiness to undertake a study of the standard at an appropriate level through a mini-course or student project, (2) progress in the standard, or (3) mastery of the standard at the desired level.

Several levels of questions would be required for the diagnostic function to address the instructional needs of spiraling for many standards. The first type of question would help teachers place students into appropriate mini-courses and provide data for guiding student work in projects. Tests using these questions would reduce the likelihood of a student trying to overcome a severe inadequacy while working on instruction that assumes adequacy. (An example would be a student having no understanding of arithmetical fractions being asked to perform statistical interpretations of data counts collected from biological field studies. However, a lead teacher might use the count data to help the student recognize the value of fractions and provide a start toward a deeper understanding of the mathematical tool.) The second type of question would provide the teacher with insights into a student's progression through an instruction unit.

Several levels of questions also would be needed for the third type, possibly to accommodate a ranking system such as described below. Students would need different questions for judging mastery of STEM content and applications for citizenry. In a detailed analysis, each student's command of each standard used in a school's curriculum would be available to the teaching team. The analysis potentially would be accessible to parents and subsequent educational institutions. Questions would be available for use in summative exams for general grades and for benchmark exams for consideration of promotions in the ranking system that follows.

Ranks

Typical grades give only a snapshot or short-term perspective of a student's performance. They do not ensure that a student commands all aspects of the expected knowledge at a specified point in a course or program. For STEM, students need to demonstrate two different sets of competencies: (1) for those with a STEM career focus, command of both theoretical content and laboratory (data collection or tactile skills); and (2) for all students, knowledge about the processes, skills, and goals of science, engineering, and technology that relate to becoming an informed and discerning citizen. The latter would also include components relevant to life skills not related to STEM alone.

The ranks implied by the entries in Table 9-1 would provide an overall general assessment of a student's level of progress toward mastery of knowledge and skills, based upon the student's performance on benchmark tests and teacher assessments of their performance as a team member and effectiveness in doing laboratory work.

At about the mid-point and end of each semester, the teacher team would use a benchmark exam to judge the general STEM status of students. The actual score to be established for entrance to a given rank would be suggested by an examinations committee of the SDC** (and perhaps adjusted by the local teacher team). Student scores would be expected to vary in a class, with some 9th-graders scoring above 11th or even 12th-graders. Some students may make little progress in the early years, then show a burst of improvement after realizing they desire a STEM career. The ranking system would demonstrate a student's overall command of STEM in a way that a simple letter or numerical course grade could not.

Level	Classification – STEM Mastery	Classification – Citizen	Benchmark Results – STEM	Benchmark Results – Citizen
		Table 9-1: A Potential Ranking System for MSS		
Entry	STEM Cadet	STEM Cadet		
1	STEM Specialist 1	STEM Citizen 1	≤D*	≤D*
2	STEM Specialist 2	STEM Citizen 2	≥C	≥C
3	STEM Specialist 3	STEM Citizen 3	≤B	≤B
4	STEM Specialist 4	STEM Citizen 4	=A	=A
5	STEM Specialist 5	STEM Citizen 5	Exemplary	Exemplary

*Numerical average as specified by the school, perhaps following SDC suggestions.

SDC would provide mechanisms for convenient evaluation and analysis of all types of student tests. Recommendations for the next instruction steps would follow the analysis as appropriate for the curriculum designed for the specific school. Further, SDC would list instructional materials that would provide support for instruction addressing the recommendations. The materials would range from print materials to support teacher instruction, to self-administered media instruction. In all situations, teachers would determine how best to proceed.

Conclusions

MSS success depends upon establishing each student's status relative to new STEM instruction and frequent assessments to ensure each student is advancing as desired. SDC would be a valuable asset for aiding the creation of assessment instruments, analyzing results, suggesting next steps for instruction, and suggesting supporting instructional materials. A student ranking system would help decision-making about future education choices and career possibilities by judging a student's current capability in STEM.

** The examinations committee should have a membership distributed equally among STEM teachers, post-secondary STEM leaders, and non-academic STEM personnel.

Using a bifurcated grading approach would recognize that STEM has two broad elements: preparation for more work in STEM, and preparedness for life and citizenship. All students must satisfy the latter element, but some may choose not to or be unable to realize success in STEM work. For students in the latter group, bifurcation offers an alternative to withdrawing from rigorous STEM instruction, and instead positions them to gain some knowledge about STEM in order to become more perceptive citizens and improve their life skills.

10

Architecture and Safety

Proposal: Provide a flexible, safe, and appropriately stocked project workspace for student teams to work with adequate supervision by teachers and communications with CSS advisors.

The following discussion relates to an ideal architectural arrangement for MSS that few schools would be able to provide in an existing structure. Fortunately, STEM teachers are resourceful and would determine how to make existing facilities accommodate some of the demands for the flexibility required by the MSS curriculum until a more suitable resource could be built. The facilities available would dictate the projects permitted for student selection. Whether new or retrofitted, the design of facilities should require input and approval from the school's STEM teachers.

Three types of physical accommodations for instructional requirements need to be satisfied for MSS: (1) lecture/recitation/study rooms for mini-courses, student team meetings, and other classroom uses; (2) work areas with broad flexibility to accommodate scientific experimentation, data collection from external sources, construction and operation of artifacts of various sizes, and communication with CSSs; and (3) one or more stock rooms. The latter includes a varied inventory to address needs for hands-on instruction across the spectrum, from scientific research to constructing artifacts at levels appropriate to high school students.

Specifications for the Ideal Architectural Setting

1. One area of the school building incorporates all the STEM instructional facilities.
2. Four or more classrooms, each with a capacity of 15 students and a teacher, are provided with room arrangements adaptable for different types of sessions.
3. All classrooms are equipped with modern audiovisual devices and communications.
4. Classrooms provide a demonstration bench with plumbing and electrical connections.
5. Study carrels to accommodate one to several students are available when the classrooms are too few or are otherwise inappropriate.
6. A large workroom/laboratory room (W/L) provides individual team areas of at least 150 ft^2 and accommodates the maximum number of students allowed in a single STEM class.
7. Each W/L team area provides an empty floor with convenient links for water/drains and electricity for maximum flexibility, plus portable laboratory benches and tables.
8. W/L includes a stockroom unit that safely provides storage for chemicals, biologicals, tools, measurement and testing equipment for sciences and engineering, and supplies.
9. W/L provides access to a materials processing (sawing, drilling, grinding, painting, etc.) room with dust and gas controls to prevent pollution of adjacent areas.
10. W/L provides the latest safety equipment and personal protective equipment (PPE).
11. An auditorium-like meeting area accommodates the maximum number of students allowed in a single STEM class (perhaps 100 students in a large school) when needed.

Overall Structure: The STEM instructional area would accommodate:

- Assembly of small classes for mini-courses
- Student team meetings
- Student project activities ranging from practicing common laboratory skills such as pH measurements to repair of internal combustion engines to the construction of model houses
- Stockrooms for chemicals, biologicals, tools, measuring apparatus, and other physical materials

Classrooms: Four or more small classrooms would enable three teachers to conduct mini-courses simultaneously with a student team meeting in the remaining classroom, while one teacher oversees all team activities in the W/L. All classrooms should facilitate instruction with current audiovisual instructional devices. They also would include communications apparatus to enable communication to external resources such as CSSs. Each classroom would have a demonstration desk with water, gas, and electrical services. Flexible seating arrangements should facilitate discussions and group problem-solving as well as lectures and testing.

Occasionally, the MSS would use a large room to accommodate as many as 100 students in a large school for organization purposes and demonstrate STEM topics' connectivity. The space need not be dedicated to STEM.

Workroom/Laboratory: The wide range of activities required for STEM instruction would require an ample open space subdivided to accommodate student project work. Each potential project area would provide for electrical, gas, and water connections. Portable work tables, laboratory benches, and rack stands would be available and located as needed for the project work. Since each student team would have as many as 10 students, space planning would provide as much as

150 ft^2 for each anticipated student team. (However, this requirement should be the specific subject of some research).

The W/L should enable one teacher to provide general oversight of the entire room for short time periods. Usually, lead teachers also would be present to help ensure safe operations and discipline; however, they may be focused on the needs of specific projects and not notice the development of an undesirable situation elsewhere in the W/L.

The periphery of the W/L would have tables or benches for a wide variety of measurement and testing apparatus of limited portability. Also, safety equipment such as fire extinguishers, adequately ventilated chemical hoods, spill containment stations, showers, and other safety devices and materials would be located for prompt and efficient use if ever needed. Personal protective equipment would be located conveniently for a wide variety of possible circumstances.

The W/L should include a containment area for projects that may produce dust and other pollutants. Also, some wall and floor space should be reserved for instructional displays and exhibiting exemplary student work.

Stockroom: Presenting STEM instruction requires ready access to many materials. The storage area for MSS must accommodate many different types of materials, tools, and equipment. Storage cabinets should be numerous and specialized. Storage for hazardous materials should have proper ventilation and electrical grounding. Safety data sheets would be made available for teachers, students, and inspectors. Many STEM organizations publish storage information relevant to their specialties for high schools.

Safety

Safety receives only brief attention here but should permeate every activity of MSS. Student accidents should be rare and of limited severity. Students would leave MSS with significantly increased safety consciousness for applications to their daily lives, future education, and work pursuits.

Safety issues would be addressed using up-to-date designs and equipment and formally incorporating safety information and procedures throughout the instructional program. Students would be encouraged to develop good habits of identifying and mitigating risks. Teachers and CSSs would teach students to identify and address inadequately controlled energy sources and other hazards, use standard devices and procedures for extinguishing fires in all types of environments, and how to clean up spills.

Each student project would include safety protocols appropriate for anticipated procedures and possible extensions or unanticipated student actions. Teachers would ensure students know about safety considerations and have access to proper materials and PPE before undertaking any hands-on activity. Students should not be afraid to do hands-on work, but should recognize that they can pursue almost any activity safely with adequate planning and proper equipment. Students would be discouraged from undertaking activities without appropriate supervision and training.

Students would be made aware of sources for safety information and devices. Instruction would make use of resources from many STEM organizations. CSSs would be encouraged to address safety issues as part of their introduction of each project in cooperation with the lead teacher. Teachers and CCSs should consider safety during the evaluation of all reports on student projects. Both the lead teachers and CSSs also should maintain close attention to safety concerns for all hands-on activities of each project.

Conclusions

Ideally, MSS would require a unique physical facility to accommodate its instructional features and maximize student learning opportunities. Classrooms for mini-courses, student team meetings, and student study need to be augmented with occasional access to a large room to accommodate the entire STEM class. An extremely flexible workroom/laboratory should accommodate activities as varied as chemical synthesis and model house construction. The design and operation of storage facilities must accommodate chemicals, biologicals, tools, other physical materials, and provide security.

Teachers and CSSs should address safety as an issue for daily life and ensure students undertake all hands-on activities safely.

KENNETH M. CHAPMAN

11

Next Steps

Having reached this chapter and decided that the time is right for developing a substantial national service for supporting high school STEM education, you may already have a plan in mind for the next step. The effort must recognize that high school education in the United States is a local effort jealously controlled by parents through elected boards of education, superintendents, and classroom teachers. Any high school STEM system like that described here must ultimately operate as a service that satisfies local needs and desires. The following discussion may be of some value if you have not already formulated a plan. Implementing your own plan will reach a higher level of success than trying to follow someone else's strategy. If you want others to join in the effort, your colleagues must be involved in the planning process.

Changing high school STEM education from a cottage industry to a comprehensive system will be an immense effort. However, the expected benefits for students and the nation's workforce will make an effort worthwhile. The MSS is only a concept. Like developing a new chemical product requiring a new process for large-scale production, several scale-up steps are necessary for MSS, with intense research at each scale, before attempting general implementation. The example of moving a chemical from an idea to a commodity production may suggest the scale-up need. A small-scale laboratory effort is required first to ensure a thorough understanding of the chemistry involved in making the chemical. Then a large-scale laboratory production model might reveal some unexpected procedural problems. If results warrant, a pilot-plant-scale effort would study the engineering requirements and perhaps produce enough product for customer evaluation. If the process is unusual and complicated, a semi-works scale-up may follow. After the project leaders believe they understand all the problems and requirements, they design a full-scale plant, direct its construction, and commence start-up. Any system like MSS must receive a comparable detailed study and scrutiny before being released for general use. These processes may take a long time.

The first requirement is proving the effectiveness of the components of the new system. Even the earliest efforts toward developing the proofs-of-concept should lead to valuable improvements to the MSS suggested in the previous chapters.

A Suggested Initial R&D Plan: The following presents an outline of a possible plan for developing a proof-of-concept for each of the MSS components and demonstrating relationships among MSS components. Potential grant developers and grantors may create better systems that differ radically from MSS.

Assume for the following "R&D plan" that MSS offers a goal for establishing a high school STEM education system. MSS components need to be considered in much greater depth than presented here and tested in realistic conditions. Researchers should re-analyze cost-benefits for each MSS component as assumptions turn into facts and become quantifiable. Costs of personnel and space would become more precise as work progresses and enable more accurate predictions. However, predicting the financial benefits of improvements in equalizing student learning opportunities, eventual workforce effects, and societal values to be gained would remain debatable; however, these are the ultimate goals of implementing a system like MSS.

The first step should be a small, invitational conference composed of STEM stakeholders: nationally recognized leaders; teachers from several different types of high schools; representatives of member organizations of STEM specialists, including scientists, engineers, and tradespeople. The conference's goals would be to determine if a drive toward making a significant change in the structure of STEM high school education should be engaged and, if so, establish a set of recommendations for the characteristics of the components of a new structure. The conference also should describe the characteristics of a General Director and Steering Committee members to lead the research effort. All members of the Steering Committee need to be committed to the idea that a high school STEM system is a desirable goal and worthy of serious study, but need not be committed initially to a specific system such as the MSS.

The following table suggests four possible initial concurrent projects. Project A might take precedence, since its findings would influence the work of the other projects.

Table 11-1: Suggestions for Four Initial Projects

Project	New Elements	Time (in years)
A	One Course Replacing Three Courses; and Teacher Team	1
B	Sample Student Projects, Related Mini-courses, and Catalyst STEM Specialist	1
C	STEM Service Centers; and Catalyst STEM Teachers	1
D	STEM Database Center	1

Post-secondary institutions and non-profit STEM organizations are potential candidates for developing and sponsoring projects. The "permanent" employer of the Director for each project should provide the physical site for offices and meetings. Project B would need constant access to media production specialists. Projects A, B, and C should work closely with nearby schools or school systems to judge practicality and provide field testing. Project D should maintain close relationships with the other projects and perhaps share some personnel.

Having reached consensus on their general missions, goals, and general operating modalities through discussions of the Steering Committee, the projects would operate independently with regular communications to the General Director and each other.

Project A should work closely with one or more local schools to develop guidelines and detailed job descriptions for teacher teams and create one or more suggested curricula for a single three-year STEM course, with an elective fourth year. Project A would create a sample list of mini-courses meeting specifications such as those indicated in Chapter 5. It also should coordinate its activities with the sample student projects created by Project B.

Project B would develop several student projects meeting specifications such as those suggested in Chapter 4. The project would recruit some CSSs for training to engage with field testing of student projects. The sample student projects should include test questions and answers for incorporation into the work of Project D. The CSSs also would cooperate with the database building of Project D.

Project C would determine the detailed characteristics of an SSC, perhaps with some variations, and make budget estimates for national operations. Several STEM teachers should be engaged in the project as potential CSTs. The CSTs might engage through Project A to model roles and fine-tune specifications for working with teacher teams.

Project D would establish detailed specifications for all components needed for an SDC and estimate budgetary requirements for development, start-up, and continuing operations. It would work with CSSs from Project B and the developers of national, state, and local STEM education standards. It also may need to work with purveyors of supporting materials for high school STEM education. Mechanisms for correlating student data from diagnostic and formative testing with a school's curriculum should suggest specific instruction for individuals and groups of students having similar instruction needs. Project D would establish procedures enabling teachers to select and to make contacts with CSSs. It would develop guidelines for presenting reviewed student projects in the database. It would cooperate with Project B to create procedures for reviewing projects.

The Steering Committee would consider the results of the projects to determine if any proof-of-concept has:

- failed and is irredeemable;
- failed and may be salvaged with modifications; or
- worked satisfactorily, and additional research should proceed.

The Steering Committee should make clear recommendations about the next steps.

As an example of the responsibilities of the various projects, consider an initial stark outline for Project B activities. Project B would develop and evaluate student projects, each requiring engaging with local schools and STEM employers. Project B needs proximity to one or more cooperating high schools with enthusiastic and competent STEM teachers and supportive administrators. It also should have good relationships for sources of project contexts in industry and research organizations. Deep connections with one or more member organizations of STEM teachers and STEM specialists would be desirable.

The broad tasks of Project B would be to:

1. Convene student project development teams that include non-academic STEM specialists, STEM teachers, and media specialists.
2. Originate contexts for student projects through work with employers.
3. Create projects and support materials for relevant mini-courses that challenge student teams appropriately.
4. Connect context originator and high school teacher(s) through a STEM expert who is deeply concerned with secondary education and committed to making structural changes (possibly the Project B director).
5. Produce for each student project unique materials for students, lead teachers, and CSSs.
6. Produce for teachers and CSSs general training materials about managing student projects and interacting with students doing projects.
7. Produce training materials for lead teachers and CSSs for each student project.
8. Guide field testing by recruiting schools as test location(s) and developing buy-in from school administrators and STEM teachers.
9. For classroom implementation of student projects and related mini-courses, make recommendations for CSS/students/lead teacher interactions.
10. Evaluate results and modify all products and recommendations appropriately for school implementation.

Implementing MSS in a High School

Public high school education in the United States is a local effort, usually guided by a Board of Education composed of residents and administered through a superintendent. Many densely populated areas have several locally controlled high schools. Most private schools have an independent Board and a School Head. Finally, each teacher meets with a small number of students, usually referred to as a class. The teacher may be supervised closely with a fixed curriculum, or given wide latitude to establish a curriculum and teaching style. Thus, MSS provision of a national-level service such as the SDC and a regional service through an SSC must avoid directives and ensure the services support local desires and requirements. These services must provide significant value to the STEM teachers and support the enhanced student outcomes desired locally.

Implementing MSS depends upon developing a desire for the proposed restructuring and augmentation change at local levels. Existing teaching personnel would need professional development for various components such as developing curricula, creating mini-courses, applying SDC tools, working with CSSs, and teamwork (working in a teaching team and guiding student project teams). Model school implementation schemes should be created during the late research phase and focus on service, not directives.

Implementation of MSS should be divided into at least two thrusts: (1) building the national and regional support elements of MSS (SDC and SSCs); and (2) efforts required at the levels of school districts and individual schools to prepare teachers and create new curricula for local implementation. Developing the details of these implementation efforts is not addressed in this book.

The Ultimate Goal – Operating at Steady-State: Steady-state operation in an educational setting is always a far less "steady" situation than a machine or process operation. Steady-state in MSS must be defined at two levels for local implementation: (1) national and regional operation that enables a school to obtain all the services of the SDC and SSC and receive full support to develop and to sustain the local curriculum and STEM teachers' assistance; and (2) for the local high school, having a teacher team with experience with MSS and some students who have reached the top level of experience within the system and can lead project team activities effectively.

National Operations: One aspect of steady-state would be achieved with a fully operational SDC, perhaps with subsidiaries due to the volume of activity. The computer database would be populated with:

- National and state high school STEM education standards.
- Data about and contact procedures for CSSs.
- Test items and programs to analyze student test results.
- Descriptions and downloadable contents of student projects.
- The curricula of individual schools.

The SDC would provide convenient opportunities for linking student test results for analysis and recommendations with a school's specific curriculum and the standards database. Teachers would find contacting and using CSSs convenient and fruitful.

SSCs would exist and employ a sufficient number of CSTs to meet the needs of the schools in their respective service areas. They also would maintain a supply of materials and equipment to support student projects and furnish services to ensure students obtain data to complete their projects. They would help STEM teachers with advice, services, and professional training.

Local High School Operations: Steady-state would be reached two to three years after enrolling the first students in the local implementation of MSS. At this point, the school would have:

- Teachers with direct experience in operating all components of the system.
- A STEM curriculum developed, tested, and modified to address local needs.
- Some students sufficiently experienced to serve as project team leaders.

Making the structural changes for MSS in high school STEM education will have to overcome extensive resistance to changing educational patterns set in tradition, modifying architecture, investing in new teaching habits, reconstructing working relationships, and creating new instructional materials. In addition:

- Some parents like to see their education repeated for their children.
- Many school administrators like to have complete control of all decision-making and the interfaces between different parties.
- Teachers invest enormous amounts of time and effort to develop the supportive components for their teaching style, and they may not desire to surrender their singular control of classes to a teaching team.
- Academically successful students may dislike surrendering their established patterns for achieving good test performance to frequent diagnostic and formative testing that pointedly reveal deficiencies.
- Academically less successful students may dislike strong encouragement from both the system and their peers to perform at high levels.
- Policy-makers may view the reduced reliance on content standardization negatively.

Overcoming these obstacles will take time, consistency, and patience by the proponents of the changes. Nonetheless, the potential benefits to teachers, students, and the workforce are enormous. However, maintaining the current structure will continue to:

- Enable too few students to achieve success in preparing for continued study in STEM or become ready for the workplace.
- Overwork teachers and decrease their availability to students.
- Deny many students from underrepresented minorities access to STEM careers.

- Limit the success of multitudes of excellent extracurricular efforts to provide students experiences in STEM endeavors.

Conclusions

After convening a small conference of stakeholders in STEM education supports reconsidering the structure of high school STEM education, research would need to address several key elements. The new model for student projects should be demonstrated through actual products and followed by classroom evaluations of the design concept, plus the documentation and preparative work needed by lead teachers and CSSs. Planning and demonstration projects would be required for both the SDC and SSCs. Evaluation of the concepts would be continuous, and many modifications of the MSS presented here should be expected.

APPENDICES

APPENDIX A - BUILDING AND MANAGING STUDENT TEAMS

"The teacher is, of course, an artist, but being an artist does not mean that they can make the profile, can shape the students. What the educator does in teaching is to make it possible for the students to become themselves." --- Paolo Freire[1]

Individualism v. Teamwork: Focusing on building teamwork expertise must in no way detract from stimulating the creativity of each student. Students should be encouraged strongly to develop their talents for independent thought and creativity. Teachers should enable each student to maximize their creativity and unique perspectives. Blending those traits effectively into a team requires students to understand how to communicate their ideas well to achieve adoption or compromise in order to improve upon the original idea. It is only by maximizing the contributions of each member that a team develops the best result.

Teamwork in the Workplace: Creativity and innovation typically are initiated by an individual often working in solitude. However, elucidation and exploitation of creation or innovation in the STEM world usually requires a team effort. Through the MSS, individual creativity should be encouraged while good teamwork habits also are developed.

Teamwork is a constant part of human existence, but it can look very different, depending on the context. The teams assembling automobiles, for example, have thousands of members working simultaneously to achieve a high degree of efficiency. Sub-teams may consist of hundreds of members to design and assemble the engines. Sub-teams of only a few people may install the windows. There is a similar dynamic at schools, where teachers, administrators, and other staff members work together to enable students to learn effectively and have a positive experience. In both examples, team members working together provide a reasonable likelihood of success; only one person working with an anti-team attitude can cause discord and create disruption.

Most people like working in teams and contributing their rightful share to the total effort. However, contributors to teams may improve their performance by practicing guided teamwork as in coached sports teams. Some people with poor teamwork habits fail both themselves and others. Patterns of good teamwork and leadership, however, can be developed or improved without discouraging independent thought and work.

Extensive research shows the open collaboration of good teams unleashes energy that boosts creativity, productivity, engagement, communication, and efficiency[2]. On the other hand, teams can also fail, such as when **they splinter into factions or divisions, causing everyone to lose.** The employees, managers, and the employer lose financially, and cohesion evaporates.

Thus, establishing and maintaining good teamwork is not just a good idea; it is essential. Further, much evidence shows that working in a team makes each member smarter, more creative, and successful.[3] The high school years offer some students their last chance to develop teamwork skills without the high cost of failure in college or the workplace. Projects in the MSS provide an ideal situation for students to learn about teamwork and build the relevant skills needed to become valuable team members.

MSS and Teamwork: The MSS expects student projects to consist of two principal thrusts: 1) teaching STEM content and developing students' skills in STEM content and processes, and 2) developing teamwork skills across the spectrum, from contributing to leading. Success in future STEM education and STEM occupations requires both thrusts.

Concurrent with STEM content and processes, the MSS contributes to building teamwork skills within a milieu of authentic contexts, clear goals, varied paths of exploration, and reporting. One of the reasons for having a three-year course is to permit student project teams to have some members with well-developed teamwork skills while others are only starting to develop such skills. Together, lead teachers, CSSs, and team leaders should help less advanced students develop analytic and robust teamwork skills and prepare for leadership positions.

STEM in high school offers excellent opportunities to build teamwork skills that will prove valuable in later pursuits. Good teamwork skills seem to be a natural asset for some people; most, however, need to work intentionally to develop such skills. Some undergraduates and many graduate STEM students need these skills to work with and manage teams conducting research.

Perspectives on MSS Student Project Teamwork: The following suggestions about teambuilding are presented in the narrow context of the MSS. Many high school students have had experiences in sports teams where good teamwork usually provides immediate rewards. STEM projects offer opportunities for students to learn that collaboration also is required for activities having a distant, and perhaps unclear, reward. An extreme example of teamwork was the Manhattan Project of World War II, which required thousands of scientists and engineers to conduct work without awareness of the explosive end objective. However, the MSS projects take only a few weeks, and enable students to identify and achieve the desired result with expert guidance from a lead teacher and a CSS.

Good "soft" skills are at the core of being a good team member. Respect for fellow team members must come first. Respect includes recognizing that other team members have different beliefs, priorities, and levels of expertise – and may be enduring unknown challenges. Good working teams diminish the negatives and emphasize the positives to reach goals as effectively and efficiently as possible. Team leaders are responsible for seeing that the intersection of technical and personal issues is symbiotic.

The developers of MSS student projects should design authentic learning experiences and challenges to advance the knowledge and skills of each team member. The lead teacher, in conjunction with team leaders, must ensure that every student is contributing and learning. CSSs should assist the efforts of lead teachers in teamwork skill development. Diagnostic and formative testing should identify whether individual students need to take one or more mini-courses or do independent self-study. Students starting the course may have more mini-course requirements than their more advanced team members, thus making them occasionally unavailable for team meetings or assignments. Advanced students may have consultations with the CSS to advance their technical knowledge. Each student should complete a project feeling they have increased their knowledge and skills. The lead teacher should ensure that leadership positions are earned and practiced effectively.

Assembling the Team: The project team would consist of a lead teacher, up to 10 students, and a CSS who would usually participate virtually. A CST from an SSC also may be a team member, to a greater or lesser extent, to ensure that adequate expertise would be available as needed by the students.

The teacher team should always select the team members. Every student team member should contribute to achieving the objectives of the project. Some members may need peer teaching. The lead teacher should ensure each team member is challenged to extend their learning and development, even when they are already very knowledgeable about the project's content.

The team members would provide a range of qualifications and capabilities in STEM knowledge and skill, from low to high. Not maintaining a constant team membership over several projects reduces the continuing effects of internal conflicts in a team that may arise from inside or outside the STEM classroom. With changing team membership, students with lower qualifications would profit from the experience of receiving peer instruction from several more highly qualified teammates. Students who may rely excessively on other team members would have fewer opportunities to do so and should be steered away from doing so by the lead teacher. Clique formation would be discouraged. Students would need to reestablish new social relationships many times during the three-year course, and take new roles in teams for each project as their expertise improves. The lead teacher would identify one or two team members to serve as the leader(s). The lead teacher also would ensure that leadership positions have been earned and are practiced effectively.

Student project team leaders should recognize their responsibility to see that the team: (1) provides engagement for every team member; (2) manages time efficiently; (3) communicates effectively both within the team and with external personnel (lead teacher, CSS, and others as appropriate); (4) behaves ethically and respectfully; (5) manages change; (6) recognizes and takes appropriate risks; (7) makes decisions; (8) displays resilience; and (9) handles conflict responsibly. Accomplishing the preceding requires student leaders to respond positively to each team member's unique motivations, abilities, biases, and experiences. The lead teacher should help student leaders learn to lead, as well as to learn technical content. These are high expectations, but the adolescent brain is still very malleable, and the soft skills developed in high school will influence each student's immediate and long-term future.

A helpful perspective for student leaders to learn is to be "servant leaders." Putting the needs of others ahead of one's desires, sharing power, and facilitating the learning and growth of team members are hallmarks of servant leaders. Achieving interdependence of team members while maintaining individualism will be difficult, but is still a necessary goal. The student leaders, the lead teacher, and the CSS may hold and promote strong opinions; however, they should remain open to alternative views and be empathetic while requiring the performance and accountability of the team and its individual members. At this stage of learner development, the lead teacher should use student leaders' frustrations and anxieties as opportunities to teach techniques to diminish and manage such stresses.

Being a servant leader does not mean doing all the work of the project. Instead, the leader should coordinate and facilitate the work of subteams and individual team members. Just as each basketball team member is expected to function at a high level throughout a game, each team member must contribute and support the team effort. To perform as effectively as possible at this level, students should become aware of the relative responsibilities of team leaders and members.

Team leaders must be vigilant and prioritize activities. It may be tempting to finish easy tasks first, but doing so may result in unproductive sequencing or interfere with higher priority or more complicated tasks. For the MSS, the leaders also have a significant role in teaching and training the less advanced students. The leaders must insist that all team members contribute to the team's efforts and achieve needed learning. They must enforce safety requirements. Team leaders will have frequent interactions with the lead teacher and the CSS and ensure that other team members' questions are addressed and the answers relayed. Student team leaders should control their emotions and frustrations and seek help with problems and crises.

Time Management: Time is a critical commodity that most high school students do not understand well. They often view time as an instant in which something is to occur. Each student's school day class starts at a specific time and ends at a later specified time. The time interval between starting and ending class times usually is perceived as a period governed by the teacher, and the student has little sense of control. However, leaders of STEM teams in the outside world recognize that time is often their most limiting factor, and failure to meet deadlines may be costly. The high school STEM classroom may not be as restrictive as the outside world, but students should be learning that time is a commodity to be managed. Although unforeseen conditions may require modifications, a timeline and deadlines are essential elements of project management.

During the planning process and throughout the plan implementation, student teams can benefit from assessing whether individual components of the project are:

- urgent and critical to the sequencing in the project.
- urgent but not critical to project success.
- neither urgent nor essential to the team, but perhaps critical to one or more team members.

These designations would help the team sequence the project components and identify the learning activities needed by the team or individual members. Project plans always are amorphous and subject to adjustments driven by operational circumstances.

Calendar Day	Action
1 (Monday)	CSS: Introduction of student team to the project
2	Initial team planning meeting; project products identified
3	CSS and lead teacher comment on plan; plan revised
4	Information and data collected as assigned Equipment needs identified and sourced Space needs identified and resolved

5	Information and data collected as assigned Team review of information collected; assignment of next steps Lead teacher and CSS review and resolve safety issues
8	Information and data collected as assigned Revise plan if necessary
9	Information and data collected as assigned Assemble presentation materials for data and information collected
10	Information and data collected as assigned Lead teacher and team leaders review operation to date
11	Information and data collected as assigned Lead teacher and team leaders review data and consider revisions to the plan and operation
12	Information and data collected as assigned Lead teacher, CSS, and team leaders review data and consider revisions to the plan and operation
15	Data analyzed and report draft initiated
16	Draft report completed

17	Draft report presented to the lead teacher
18	Final report completed and submitted to the lead teacher and CSS
21	Responses are given to the team by the lead teacher and CSS

Figure A1: A generic model spreadsheet for project timing for a 3-week project

Maintaining a spreadsheet should be suggested to team leaders to focus the use of time and personnel assignments. The student team should outline an overall project plan based on time, member assignments, and activity sequencing. Activity relationships should be established and adjusted to meet reality as the project proceeds. The generic spreadsheet in Figure A1 for a three-week-long project is only partially complete. The actual team spreadsheet would have a column for each member with their assignments, progress, and other notations. As the reality of data and information collection becomes evident, the team should add extra days as required. The lead teacher would emphasize overlapping activity possibilities and promote efficient use of time.

Communications: Two-way flow of comments, information, and opinions is necessary for the effective operation of teams and transmission of all helpful information. Vocabulary, abbreviations, and communication characteristics should be dictated by the least competent recipient's knowledge and skill status, not the originator's preferences. The tone should be neither aggressive nor accusatory, but assertive. Problems and differences need resolution, not blame assessment. Conversations need to be respectful, with all parties allowing or encouraging full disclosure. Lead teachers should ensure that team members avoid sarcasm and build trust so that both negative and positive issues can be discussed thoroughly.

Conflict: Lead teachers should redirect disagreements to uncover alternative pathways for solving problems and resolving issues. The lead teacher and team leaders should recognize that some differences in beliefs and values may require that certain topics be divided to prevent them from jeopardizing the mission. Mitigating discord may require intervention by the lead teacher, possibly through team meetings or with the protagonists in a non-threatening environment. The focus should be on issues, not people, and efforts may be needed to modify undesirable behaviors. Even though the students are assigned responsibilities, the lead teacher is still the ultimate responsible party.

For each student project, a statement by Leon C. Megginson *in 1963 may be helpful as guidance for the lead teacher: "According to Darwin's Origin of Species, it is not the most intellectual of the species that survives; it is not the strongest that survives; but the species that survives is the one that is able best to adapt and adjust to the changing environment in which it finds itself.*[4]*"*

Key resources used in developing this appendix include:

- Hess, D. Transitioning to a Leadership Role. *CEP*, January 2020, p. 46-49.
- Hess, D. *Leadership for Engineers and Scientists: Professional Skills Needed to Survive in a Changing World*; John Wiley: New York, 2018.
- Sadoway, D. Introduction to Solid State Chemistry, Lecture 26: Acid-Base Chemistry. MIT OpenCourseWare Course 3.091.
- Drennan, C. Principles of Chemical Science: 5 Lectures on Acid-Base Titrations. MIT OpenCourseWare Course 5.111.

References

1.   https://www.goodreads.com/quotes/8776506-the-teacher-is-of-course-an-artist-but-being-an (accessed Dec 21, 2021).
2.   Middleton, T. The importance of teamwork (as proven by science). Blog post on Atlassian website, Dec 15, 2019. https://www.atlassian.com/blog/teamwork/the-importance-of-teamwork (accessed Dec 21, 2021).
3.   Shannon, A. "What do you lose when team-work fails?" Blog post on Kent State University's Center for Corporate and Professional Development web-site. https://www.kent.edu/yourtrainingpartner/what-do-you-lose-when-teamwork-fails? (accessed Dec 21, 2021).
4.   https://www.darwinproject.ac.uk/people/about-darwin/six-things-darwin-never-said/evolution-misquotation (accessed 01/16/2021)

To the Student: Planning was a significant factor in getting your last sandwich to you, whether meat or vegetable. If it was a plant-based burger, farmers in Iowa or Brazil had to plan and arrange to grow soybeans and other vegetables and grains for the contents. After the farm products were transported to a processing site, cleaned, and concentrated, a nutritionist selected a recipe to make a tasty and healthful patty, and the product was manufactured. Transportation specialists arranged for trucks, railroads, and airplanes to move the burger from its manufacturing site to a restaurant or store near you. The burger was cooked, perhaps by you, for delivery to the location where you consumed it. Many people, including yourself, engaged in both short-term and long-range planning just to give you a veggie burger. Everyone participates in planning, often unconsciously.

A planning process provides the structure for the scientific method and engineering design. Initial planning clarifies a product, process, or report that constitutes the conclusion of the project. The plan lays out the strategy, assignments, budgets, and timeline for arriving at the end. (For the student projects in MSS, budgets usually are not required but may sometimes be included for instructional purposes.) As implementation proceeds, new information, data, and realities may allow succeeding steps to proceed as planned, and it is common for problems to arise that spur feedback and demand modifications to the plan – sometimes as far back as its initially identified objectives.

Planning may be an effort by a single person. In early 1941, Charles Sorensen, chief engineer for Ford Motor Company, took part in a one-day tour of the manufacturing facility completing the assembly of one B-24 bomber made of about 0.25 million parts in three days. Using the information he gathered, he developed a production plan overnight which enabled his company to reach production rates of one B-24 bomber of 2.2 million parts every 63 minutes.

Planning is frequently a team effort. A famous example of this was the Manhattan Project that led to the development of the atomic bomb. The project required multitudes of plans by individuals and teams. The project eventually employed approximately 130,000 people, and required the contributions of nearly one-third of all US citizens trained or training in STEM fields, from Nobel prize winners to undergraduate students.

Most STEM jobs require conscious planning, and many organizations have developed planning procedures specifically tailored to their needs. The components of planning introduced here are typical and use words and techniques that may differ from those you will encounter elsewhere. For maximum utility, planning should be purposeful and organized to fit the need – whether that need is to build highways, move military units, or make dinner for your family. The planning skills you learn in your STEM course will apply to any planning process you may use in the future.

Developing and changing plans, whether consciously or unconsciously, is something humans do all the time. Sometimes the planning stratagem is effective; sometimes, it is not. STEM projects provide an excellent opportunity for you to develop more organized methods of planning that are useful in STEM work and for many other human activities.

To the Lead Teacher: The planning stratagem presented here is a blending of many planning structures. Examples of some of these guides include: 1) the "scientific method" taught to many generations in high school classrooms as the process by which science is practiced; 2) the strategic planning model developed by the non-profit American Chemical Society to engage members, officers, and staff in creating long-term plans; and (3) the Design Thinking approach developed at the Stanford University School of Engineering.[1] The key idea is to get the students to understand that effective planning requires an organized way of getting from a current or initial set of conditions to a clear objective. Students should come to realize that plans also may be modified when additional information becomes available.

You are encouraged to examine many planning processes and select the one, perhaps with modifications, that will most benefit your students; alternately, you could use your own design. Planning is a continuous process, always subject to change in response to unanticipated situations and improved information. For example, a student may have planned to use mayonnaise on their burger, but catsup might be substituted. A building contractor may be using architectural plans that specify fasteners for steel beams that prove to be unavailable. Changing the fastener may require negotiations with the architect, city zoning departments, and vendors. The change may even require some unexpected training for the workers who will do the installations. Plans must change to match reality. In a team operation, planning should involve the entire team at critical points, while involving only specific individuals or groups for more detailed planning.

In Figure B1, the team is engaged as a single unit at the centerline of the convergence points, and the diamonds may be traversed several times during a project in response to feedback. Even the project's original objectives may need revision as more information becomes available and reality forces changes. During divergence and convergence, individual team members or small groups may plan and work independently to advance the project plan.

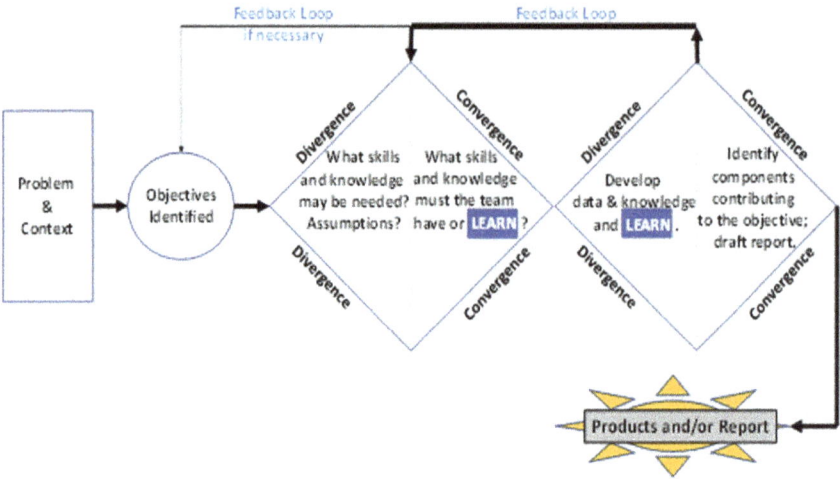

Figure B1: An Example of a Planning Process Emphasizing Learning Needs

---

Specifications for Project Planning and Management

1. Ensure each team member understands the goal of the project.
2. Encourage each team member to contribute to planning.
3. Identify and organize action steps.
4. Identify the learning needs of team members.
5. Strive for consensus; accept majority votes when necessary.
6. Establish responsibilities and make assignments.
7. Prioritize actions and establish timelines.
8. Ensure safety in plan implementation.
9. Review progress.
10. Adjust plans.
11. Accept failure as a learning tool.

The Components of a Planning Process: Do not expect that the steps always will be evident during implementation. Also, steps may overlap or change sequence. The lead teacher should ensure each team member clearly understands other students' work throughout the planning and reporting sessions. A useful acronym for the steps in this specific planning process is **ISCIPER**.

Identify: This step may be clear from initial communications. Project materials sometimes may provide goals and constraints directly in the introduction to the project. However, extended team discussion may still be necessary to ensure all members understand what is required.

Developers of MSS student projects may design some tasks to have goals that are vague or imprecise. The team is then required to clarify the goals. Each team member should contribute to identifying at least one goal by presenting a written statement on a sticky note for posting to be examined by all team members. The team leader(s) have the duty of writing a "final" statement of goals based upon these inputs, which the lead teacher and the CSS must then evaluate. Students in their first semester in the course may be excused from contributing written notes.

Specify: Early planning sessions should define one or more issues to be addressed or problems to be solved. The same procedure used in the "Identify" step may be employed. Specific issues may also be considered objectives that individuals or groups may be assigned to work out.

Conceptualize: The student team should brainstorm to determine the activities that may solve the problems. Activities may include performing experiments or otherwise collecting data, building artifacts, collecting information from sources, doing calculations, and learning new knowledge and skills required for problem-solving for the project. The team should develop a schedule, which then must be approved by the lead teacher and CSS.

The lead teacher and the team must identify needs for peer and teacher instruction. The team may need diagnostic and formative testing. Guided by the lead teacher, the team develops detailed plans, sequences activities, and prioritizes the work based on the identified information. Individuals and groups receive assignments and begin the work, including learning, assigned to them.

Inform: Team members are obligated to keep the team aware of the status of the work assigned to them. The team leaders must check information acquisitions frequently and call for team meetings when necessary. Team members should exert maximum effort to maintain the schedule, which may be revised only with permission by the lead teacher.

Perfect (verb): Team members experiment, build artifacts, collect information, and otherwise proceed to complete assignments. Improvement of procedures and skills may require repetitions. This step includes the instruction requirements. With the experience gained, the schedule may require revision; the conceptualization and previously identified steps may require reconsideration. The team should revise plans as experience and information dictate only with the permission of the lead teacher.

Evaluate: Concurrent with the "Perfect" step, the team should consider all data and information for accuracy and sufficiency. The lead teacher and CSS should receive interim reports, which may prompt issuing requirements for additional work, assignment changes, and additional instruction. All parties must recognize that the ultimate objective is for each team member to learn more STEM knowledge and skills.

Report: The lead teacher and CSS (client/consultant) must require draft and final reports. Revisions of drafts usually will be necessary, and only the final report should receive a grade. All team members should receive the same grade for the final report.

Identifying what each team member needs to learn is critical in a technological world where change and information advances are almost constant. Some necessary learnings are everlasting and may or may not be fun, such as developing skills in mathematics. Some learnings, such as gaming, will be entertaining and fun. Some necessary learnings are for short-term advantage and may need future adjustment, such as learning a computer programming language.

In school and some other settings, much learning meets goals set by others. That may be boring but necessary. After leaving school, many people discover that a new and interesting skill may increase salary. That may eliminate boredom.

Some team members will need to improve their skills to match more advanced team members for many team projects. Some team members may learn unique skills to supplement the skills of the other team members and help the team achieve its goal.

Acronyms, Definitions, and Perspectives

SOP: Standard Operating Procedure: A description of the specific actions or steps for doing a task or job. In science laboratories, these procedures should give results that can be compared precisely with data from previous efforts or compared with future results. The SOPs may provide results to use in conjunction with the similar work of many other scientists/technologists. Typically, SOPs are written or presented digitally and are to be followed exactly. Appropriate officials must approve any changes.

Teams and Teamwork: The lonely scientist toiling away in an isolated lab has been a rarity for many decades. Today's scientific and engineering work typically requires teams whose members individually bring appropriate expertise to a project. Some individuals may stay with a project to its conclusion; others may make a singular contribution and leave for other work. Each team member must be supportive, even when addressing failures or non-productive activities. Often, a team member may find that their contribution requires them to rapidly and efficiently learn new knowledge/skills.

QC: Quality Control: Providing customers with high quality every time is necessary to retain customers. Extreme measures are often taken to ensure consistent quality. For example, soft drink manufacturers purify the water used in their products to provide safety and ensure their product has the same taste every time. The feeds used for alpacas and llamas necessarily contain copper concentrations that are toxic to sheep and goats. Thus, feed manufacturers often place their processing of specific products in different cities to avoid any chance of contamination.

Two Examples for Student Team Planning and Managing a Project in MSS Operations

Project 1: Determining and Reporting the pH of Water Sources

The following planning and management outline is based on a hypothetical student project intended to determine and report the pH of environmental water sources safely accessible to the students. The project introduces novice students to pH and elementary measurement techniques. It also challenges advanced students to address higher-level issues of pH and chemical equilibrium.

After project completion, the lead teacher may wish to add the student data to a national database for similar projects from other schools. The consolidated data might be available for subsequent student projects exploring relationships between the original local data and scientific, sociological, and economic factors. See Table B1.

Table B1: Planning Steps for Project 1

1	Identify	Use the introduction to identify the goals of the project and the nature of the products needed to satisfy the client (CSS). (Use a sticky note session to collect the ideas about goals, steps, and products from team members. This activity should be followed immediately by an organization and consolidation of the notes. The CSS and lead teacher must be satisfied with the statement of goals/objectives before the team can proceed.)
2	Specify	Specify the problems that must be solved and clarify them sufficiently so that work and learning tasks can be assigned to individuals or groups. (This may require researching, reporting, analyzing, and studying by team members.)
3	Conceptualize	Move from the current status of the team's knowledge and skills for solving the problems that enable reaching the goals. For example, the team may agree that using a pH meter to identify specified endpoints will require: ·Calculations to convert pH to acid/base concentrations ·Algebraic skills to address logarithms ·Graphing logarithmic information ·Research, work, analyses, and reports by individuals and groups to implement the procedures identified ·Identifying the apparatus and equipment needed ·Acquiring the skills required to use the devices correctly ·Calibrating instruments and apparatus ·Maintaining consistent conditions ·Collecting data, making models, and drawing designs as necessary
4	Inform	Inform the team about the information, observations, and data collected by individuals and sub-teams. The team provides feedback to individuals and sub-teams.
5	Perfect (verb)	Perfect procedures to collect more accurate data and information.

| 6 | Evaluate | Evaluation has been used extensively in the preceding steps to judge data and information. At a point in the project, the data and information should reach a quality and quantity that enables satisfactory solutions of problems and completion of the product required. |
| 7 | Report | Write a draft for team approval and initial review by the lead teacher; present a final report to the lead teacher and CSS; and receive feedback. |

Project 2: Standard Operating Procedures for pH Titration

Abbreviated Project Description: A manufacturer of heavy-duty industrial cleaning products uses large quantities of commercial sodium carbonate (soda ash), most of which arrives in 230,000-lb lots shipped in covered (watertight) railroad hopper cars. The hopper cars are placed on a siding over an auger arrangement that moves the soda ash as it falls from the railroad car to bulk storage in the plant. The manufacturer's procedures allow only 24 hours for receiving, unloading, and returning the railroad car to avoid extra costs (demurrage).

This project aims to develop a written testing protocol to ensure the soda ash in the railroad hopper car meets the manufacturer's specifications before unloading. The protocol requires both a standard wet pH titration (with indicator) analysis and an instrumental analysis using a pH meter. The protocol must address each step and establish a typical chain-of-custody, from collecting the sample before unloading, to providing formal approval, to starting to unload the complete load. (See a routine unloading procedure at https://www.youtube.com/watch?v=02mwgh2WvPk [accessed Dec 21, 2021].) The protocol should be written with sufficient detail so that laboratory technicians could perform the procedure without prior experience with this specific analysis, assuming they are familiar with titration and using a pH meter.

Manufacturer's specification: Twenty-five milliliters of a 0.10M solution of the soda ash sample must be neutralized with 2.5mmol of HCl with a permitted range of ±0.05mmol.

Among the items to be discussed in the first planning session, after the objective is clarified, are: What are the STEM knowledge and skills required to produce the necessary data? (Assess the value of the flow of conversation/debate throughout the discussion; encourage students to identify inadequacies of their communicating.)

Key steps in the process are:

A. Collect the sample

- Sample size
- Number of samples taken from a railroad hopper car
- From where in the car should the sample be taken? Is it necessary to take samples only from the suggested locations?
- Who should take the samples?
- Safety requirements? Personnel limitations?
- What information should be provided with the sample?
- What are the requirements for gaining access to the cargo?

B. Analysis of the sample

- Solution preparation for analysis
- Indicator for titration
- Apparatus to be used for titration
- Instructions for performing the titration and identifying the endpoint
- Information about the accuracy requirements
- Instructions accompany any calculations relevant to these titrations

C. Follow-up and report

- Instructions about what to do when the sample meets specifications
- Instructions about what to do when the sample does not meet specifications

Populating the appropriate cells in a spreadsheet, such as suggested in Table B2, will help manage the assignments and specify the knowledge and hands-on skills to be learned or improved. The students needing to acquire knowledge and hands-on skills would be identified through diagnostic and formative testing. Project developers should help the lead teacher by providing a sample spreadsheet with the columns of "Project Needs," "Concepts Addressed," "Knowledge Required," "Skills Required," and "Mini-courses Offered" already completed. The lead teacher then would use the spreadsheet entries to guide the initial planning discussion.

Table B2: Sample Spreadsheet for Project 2

Project Title: SOP for Soda Ash Raw Material Analysis

Lead Teacher: Jane Doe **Student Project Team Leaders:** Amani and Diego

Team Members: Amani, Diego, Joe, Diamond, Ahmad, Sam, Shanice, Tomás, Deshawn, Santiago

Project Need	Concepts Addressed	Assigned	Knowledge Required	Skills Required	Mini-course Offered	Peer Teaching	Learner List
Sample collection	Representative solids sample collection from rail cars	Amani Joe	Statistics, basic	Safety around rail cars	Concepts of sampling	Led by Diego	All students

Table B2 addresses only the initial student team planning meeting for this project. New students may need some coaching to understand the planning process and become valuable contributors to discussions. Students familiar with MSS should be comfortable in planning meetings and contribute effectively. Subsequent planning/management meetings should be short and address specifically identified issues.

Suggestions for the Lead Teacher: Keep self-stick note pads available in 3x5 in. and larger sizes. A good and cheap substitute is paper cut to various note sizes that can be mounted with masking tape.

1. Populate the student team with up to 10 individuals with varied capabilities.
2. Obtain the services of a CSS for the introduction and first planning meeting. The CSS will add authentication you cannot, and also should provide unique career information.
3. Act in an advisor/supervisor manner and enable the student team leaders to manage the meeting. Use leading questions as necessary and give the students time to respond.
4. Begin the planning meeting by ensuring all team members are clear about the objective (a report that describes a standard operating procedure or SOP).
5. Stimulate discussion about what sections should be included in the report. All suggestions should be written on sticky notes and placed on a discussion board. Suggestions may include: Sampling, Describing the Chain-of-Custody, Specifications, Acid/Base Chemistry (as it applies to this activity), Solutions (making, maintaining, calculating concentrations, and using standards), Equipment and Data Recording, Actions After Completing the Analysis, and Record-Keeping.
6. Let students arrange the sections in a proper sequence, then critique the result.
7. Show a video of unloading a railroad hopper car, such as at https://www.youtube.com/watch?v=02mwgh2WvPk (accessed Dec 21, 2021).
8. Show a video about industrial quality control, such as the video at https://www.youtube.com/watch?v=yhZ6Mg3FQSI (accessed Dec 21, 2021). This video discusses personnel as well as quality control issues.
9. If time permits on Day 1, discuss in detail a project component such as Sampling. Discuss what would be considered an adequate sampling of a railroad hopper car for a bulk chemical such as soda ash. Engage the team in identifying the components of the

sampling activity. Have students make their contributions on sticky notes and post them.

10. Stimulate discussion of the steps for chain-of-custody from collecting the sample to getting the sample logged into the laboratory where the analysis will be performed. Use sticky notes to identify and describe each step. The team can amend the plan in the future to address overlooked steps or provide more detailed information.

11. Stimulate discussion about each proposed report section using the thoroughness suggested by Item 9. Identify knowledge and skills that need to be acquired by each student. Determine information needs that one or more student assignments must address.

12. Involve all the team members in the planning discussion; elicit volunteers for the project components and make assignments as necessary. Do not allow an individual or small group to dominate the debate.

13. Do not dispose of any suggestions; divide the suggestions into groups of principal sections and subordinate topics. Dispose of duplicates and those suggestions that now appear inappropriate.

14. Let students plan their sampling process. After discussing the size of the sample, tell them that samples could be collected in a pint-size plastic container with a lid that seals tightly. Have the team determine the information that should be written on the label of the sample container.

15. The lead teacher may wish to provide samples to mimic those that could have come from a hopper car.

For alternative views of planning procedures and engineering design processes, consider the following:

National Center for Engineering and Technology Education (NCETE) researchers built on a model framework from Massachusetts and the UTEACH design model (University of Texas) to emphasize the nonlinear nature of engineering design. This model emphasizes how designers may jump back and forth between some steps and possibly skip others entirely.

A detailed discussion of incorporating engineering design into high school STEM courses is available.[2]

The Massachusetts Department of Education is credited with producing the first state-level standards that included engineering-related learning expectations for K-12 students. Their model of the engineering design process had eight steps in a repeating loop. A description of this model is presented in the Massachusetts Science and Technology/Engineering Curriculum Framework, available online at that state's Department of Education website[3]. A version of the engineering design process is presented in 2016 Science and Technology/Engineering Curriculum Framework.[4]

University of Texas UTeachEngineering Engineering Design Process in AC 2012-4130: A Unique Approach to Characterizing the Engineering Design Process, Ms. Lisa Guerra, Dr. David T. Allen, University of Texas, Austin, Dr. Richard H. Crawford, University of Texas, Austin, Ms. Cheryl Farmer, UTeachEngineering © American Society for Engineering Education, 2012 Complete paper provided at

https://monolith.asee.org/public/conferences/8/papers/4130/view (accessed 02/01/2022)

References

1. Stanford University, Hasso Platner Institute of Design. "Get Started with Design Thinking" web page (accessed Dec 21, 2021).
2. Householder, D.L; Hailey, C.E. "Incorporating Engineering Design Challenges into STEM Courses." https://files.eric.ed.gov/fulltext/ED537386.pdf (accessed Dec 21, 2021).
3. Massachusetts Department of Elementary and Secondary Education. Massachusetts Learning Standards page. https://www.doe.mass.edu/frameworks/ (accessed Dec 21, 2021).

4. Massachusetts Curriculum Framework – 2016. "Science and Technology/Engineering, Grades Pre-Kindergarten to 12." https://www.doe.mass.edu/frameworks/scitech/2016-04.pdf (accessed Dec 21, 2021).

APPENDIX C - DEVELOPING THE MODEL STEM SYSTEM (MSS)

Just before the holiday season at the end of 2015, an invitation to a meeting for discussions about curriculum changes and a new building arrived from the parent of two of my high school students. The parent was a retired Navy officer who now managed several teams at a Navy research facility, with personnel ranging from custodians to Ph.D. research scientists and engineers. We had discussed technical workforce issues several times from her perspectives, both as a parent and as a manager of teams. She was disturbed by the U.S. education system from the perspectives of both roles.

Being a skeptic about meetings concerning changing education, I suspected that little discussion would focus on curricula, about which most people complained without offering ideas for substantial improvements, and that building issues would occupy far more time. However, I accepted the invitation.

Then, I spent a couple of vacation days trying to answer the question, "What would I want for high school STEM education if I could start with a clean slate?" I considered my experiences:

1. Six years of teaching college courses receiving students directly from high school.
2. Three years of work in industrial chemistry laboratories and designing new production equipment for polyethylene film.
3. Thirty-one years of work on post-secondary issues for the American Chemical Society, where I enjoyed a front-row seat to observe attempts to improve high school science education, including

developing the original national science education standards. I studied statistics and prepared reports about chemistry education for committees, discussed shortcomings of education with industrial representatives and educators, observed the efforts of outstanding science educators of international renown trying to improve high school science education, and worked closely with many of the latter. After 31 years, we seemed to be in the same place as when we started, although educational media had improved.

4. After retiring, I consulted with two-year college programs for several years and continued to study the alarming science educational statistics.

5. In the spring of 2006, a private school attended by some of my grandchildren called me to request help with their chemistry course. A few minutes into a visit demonstrated that the school had neither a coherent science program nor a qualified chemistry teacher. I accepted an offer to become the school's chemistry teacher. That request soon expanded to include physics. Just before the academic year began in a new building on a 139-acre campus, the school administration requested that I also teach biology and middle school physical science. While the "laboratory" in the new building was designed poorly, my request for equipping the space was fully funded, as were subsequent funding requests for supplies and apparatus. An administrator delivered to me a small cardboard box containing all the equipment, chemicals, and other supplies accumulated over the 18-year life of the school. (When I retired from the school after ten years, a 26 x 30-foot stockroom/prep room was required for the accumulated STEM equipment and supplies, and each of three specialized science classrooms had additional equipment and supplies.)

6. At the end of 2015, I had accumulated 8.5 years of experience as a high school science teacher in a nearly ideal traditional teaching situation. I had changed textbooks three times for the chemistry

class. For AP chemistry, I used a popular text, which I had re-viewed and made suggestions for presentation changes after the first edition became available some 38 years earlier. Most of my suggested changes had been adopted, and the textbook was now part of an excellent teaching "suite" presented by a major pub-lisher. Yet, I was unhappy with what I had been able to accom-plish for my students, a substantial percentage of whom pursued STEM careers.

7. I had become sensitized to many problems that seemed intractable due to the present structure of science teaching (see Chapter 1).

Then, I applied my engineer's perspective to the problems and used these steps of a problem-solving strategy:

1. Identify and define the problems.
2. Specify the requirements and the objectives.
3. Conduct background research (read the history of science edu-cation).
4. Identify the fundamental assumptions.
5. Correlate the flows (of personnel) in the system.
6. Identify the available assets – both traditional and potential – and consider relationships between subsystems.
7. Design possible solutions
   Propose a design
   Evaluate the design
   Try to determine how to mitigate problem components
   Reiterate until the optimal design can be identified
8. Create a prototype of the design.
9. Advocate for research to test the design and redesign as necessary.
10. Communicate with stakeholders (this book is a starting point).

The result was the Model STEM System (MSS) design described in this book. Steps 7 through 9 of the problem-solving strategy require extensive resources and time not available currently to me. Thus, the design must be considered provisional and subject to improvement by the appropriate personnel. For example, funding agencies could support projects located in university STEM education departments.

Continual reviews of STEM education literature and participation in conferences have failed to suggest significant changes to MSS to me. An early draft of the MSS was submitted to an NSF Big Idea program for 2018-2019 and placed in the top 1/8th of the offerings.

Troubling Observations

The need for an MSS results from two sets of observations: (1) the many unproductive attempts over 50 years by dedicated and brilliant science and engineering educators to improve the capabilities in STEM for the graduates of American high schools, and (2) the work required of technical personnel in various pursuits, from primary processing of raw materials to exploration at the cutting edges of science.

The chasm between the results of the first and the second set of observations is widening, even as the demands and opportunities of STEM employment increase dramatically. The chasm also means that many students spend more time and money than necessary for colleges, while shortening their work lives. Some potentials of modern technology that would benefit teachers are ignored, while many applications are applied somewhat haphazardly.

The MSS is offered as a starting point for discussion and action to ease the burden on STEM teachers and make essential changes in the most critical area of STEM education: the three high school foundation courses of biology, chemistry, and physics.

Nothing in this book should be construed as a criticism of the outstanding efforts of STEM teachers and administrators, foundations and membership organizations supporting change in STEM education, and the many scientists and engineers, who often with the support of their employers, are trying to improve high school education. Indeed, some innovators have created programs that include certain characteristics and components of MSS.

The Teachers' Environment

Most high school STEM educators seek to improve their instruction continuously. They take university courses, travel to locations where science is taking place, follow recent STEM developments to provide currency to daily classes, and read extensively for STEM content and pedagogical improvements. They determine how to best implement in their classrooms the recommendations of experts. These educators are besieged with demands and guidance from politicians, nonprofit organizations, publishing companies, advocates for instructional devices, leaders of the most recent education fads, and bloggers. Locally, they are driven to "do more" by state education standards, school administrators, school boards, PTAs, and parents. Individual teachers work with students of diverse capabilities, varied parental support, multifarious interests, unequal economic situations, local sociologic divisions, and sundry educational histories. These teachers, or miracle workers, try to use all the external influences and their common sense as they strive to make classes enjoyable and to enable each student to move forward in understanding the natural and technological worlds while laying the groundwork of many students for a wide variety of careers in the modern workforce.

Most of these teachers work within a structural framework institutionalized over a century ago, when Newtonian physics was applied extensively to change the way humans lived and worked. At that time, horses and steam-driven trains still provided fast transportation, even as Henry Ford was tinkering with his first automobile design. Soon, employers needed many people for assembly lines and other jobs requiring little training to do a single operation repeatedly. Academicians saw the usefulness, if not necessity, of dividing learning into single concepts and skills to facilitate learning. Educators separated natural science into courses of biology, chemistry, and physics. Pre-college institutions ignored engineering, although it was the primary focus of many colleges; additionally, some schools provided vocational training in response to local needs.

Integration of Disciplines

In recent years, academics have noted that most research and development work requires the integration of many disciplines. The term STEM (for science, technology, engineering, and mathematics) has evolved over a generation and applies to various connotations. Robots already do many repetitive functions that previously required human activity. Humans now need to integrate knowledge and skills more fully from many STEM areas through mental and physical processes. The teamwork of small groups of individuals with diverse expertise is now incredibly beneficial to creativity, problem-solving, and the advancement of goals. Living in a technological society requires every citizen to exercise judgments about daily activities that involve STEM issues, such as health and environment, and participate in civic decision-making.

Mismatch Between the Inside and Outside Realities: STEM teachers work mostly alone to present courses in a silo-like context, even in large schools (see Figure D1), teaching various STEM classes as though they were unrelated to each other. Teachers in smaller schools have few or no STEM colleagues with whom to share problems and experiences. Keeping STEM courses current requires teachers to invest time to achieve the desired depth of knowledge in both existing and new technology fields while also updating teaching skills and curricula.

Students are well aware of the rapidity of technological advancement and enthusiastically work to develop new knowledge and skills outside academic classrooms. They also are well aware that highly publicized advances in STEM fields may be completely obsolete before they are incorporated into science curricula. However, many students still avoid STEM courses as soon as possible, and do not achieve a broad knowledge of the diverse field.

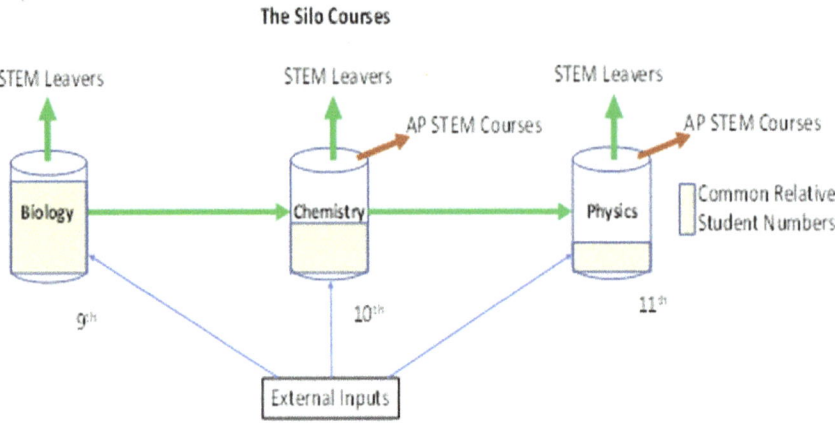

Figure D1: The Silo Effect of Traditional STEM Courses

The MSS was developed to address many of the issues implied above, assist teachers directly, ensure high school students start experiencing a broad acquaintance with STEM in their first classes, and enable students to develop "soft skills" useful in all careers.

Kenneth M. Chapman

Education
1958 A.A.S. in Chemical Technology   Ohio College of Applied Science
(merged into University of Cincinnati)
1961 B. S. in Chemical Engineering   Massachusetts Institute
of Technology
1969 M.S. in Education   The George Washington University

Employment
1956-1959 The DuBois Chemical Company, Cincinnati, OH
1960-1963 Mehle Manufacturing Company, Williamstown, KY
1961-1963 Ohio College of Applied Science (merged into
University of Cincinnati)
1963-1967 Temple University, Philadelphia, PA
1967-1999 American Chemical Society, Washington, DC
1972-1973 Computer-Based Instructional Systems, San Antonio, TX
1998- 2008 Consultant to many colleges and to the Advanced
Technology Environmental Education Center,
Davenport, IA
2006-2016 The Carmel School, Ruther Glen, VA

Publications
Pecsok, Robert and Chapman, Kenneth, Co-Editors, *Modern Chemical Technology,* 7 Volumes, Ed. 1, 1970-1972
Pecsok, Robert, Chapman, Kenneth, and Ponder, Wade, *Handbook for Chemical Technicians*, 1972
Hofstade4r, Robert and Chapman, Kenneth, *Skill Standards for the Chemical Process Industries*, 1997

Producer for the ACS of many media-based continuing education units for chemists and chemical engineers.

Served on governing bodies for:
Triangle Coalition for Science and Technology Education (national)
National Vocational and Technical Honor Society (national)
Advanced Technology Environmental Education Center (national)
American Chemical Society Virginia Section (regional)
Caroline County Agricultural Fair Board of Trustees (local)

www.ingramcontent.com/pod-product-compliance
Lightning Source LLC
Chambersburg PA
CBHW051621120626
46551CB00014B/1890